# Mechatronics Matters

## Englisch für Mechatronik

von
Kenneth Thomson
Isobel Williams

W0412017

Cornelsen

**Mechatronics Matters** wurde geplant und entwickelt von der Redaktion Moderne Fremdsprachen des Cornelsen Verlags, Berlin.

| | |
|---|---|
| Verfasser/in: | Kenneth Thomson, Berlin<br>Isobel Williams, Berlin |
| Berater/in: | Klaus Goldmann, Rastatt<br>Sibylle Weiß, Willich |
| Projektleitung: | Jim Austin |
| Verlagsredaktion: | Andreas Goebel |
| Außenredaktion: | Isobel Williams, Berlin |
| Redaktionelle Mitarbeit: | Kate O'Brien, Christine House, Fritz Preuß (Wörterverzeichnisse) |
| Bildredaktion: | Jürgen Frey, Gertha Maly |
| Layout und technische Umsetzung: | Oxford Designers & Illustrators |
| Umschlaggestaltung: | Katrin Nehm |
| Titelfoto: | DLR, Institute of Robotics and Mechatronics |
| Illustrationen: | Oxford Designers & Illustrators |

Erhältlich sind auch:
Audio-CD
Handreichungen zum Unterricht

**www.cornelsen.de**

Die Internet-Adressen und -Dateien, die in diesem Lehrwerk angegeben sind, wurden vor Drucklegung geprüft. Der Verlag übernimmt keine Gewähr für die Aktualität und den Inhalt dieser Adressen und Dateien oder solcher, die mit ihnen verlinkt sind.

Dieses Werk berücksichtigt die Regeln der reformierten Rechtschreibung und Zeichensetzung.

1. Auflage, 1. Druck 2006/06

Alle Drucke dieser Auflage sind inhaltlich unverändert und können im Unterricht nebeneinander verwendet werden.

© 2006 Cornelsen Verlag, Berlin

Das Werk und seine Teile sind urheberrechtlich geschützt. Jede Nutzung in anderen als den gesetzlich zugelassenen Fällen bedarf der vorherigen schriftlichen Einwilligung des Verlages. Hinweis zu § 52 a UrhG: Weder das Werk noch seine Teile dürfen ohne eine solche Einwilligung eingescannt und in ein Netzwerk eingestellt werden. Dies gilt auch für Intranets von Schulen und sonstigen Bildungseinrichtungen.

Druck: CS-Druck CornelsenStürtz, Berlin

ISBN-13: 978-3-464-20799-4
ISBN-10: 3-464-20799-4

Inhalt gedruckt auf säurefreiem Papier aus nachhaltiger Forstwirtschaft.

**Mechatronics Matters** ist ein völlig neu entwickeltes Englisch-Lehrwerk für Mechatroniker/innen, die am Berufsschulunterricht oder an betrieblichen Fortbildungsmaßnahmen teilnehmen. Das Werk basiert auf den Lehrplänen der Bundesländer für Berufsschulen und bereitet auf die KMK-Zertifikatsprüfung Stufe II vor (Niveau B1 des Gemeinsamen europäischen Referenzrahmens).

**Mechatronics Matters** richtet sich an Lernende, deren Englischkenntnisse dem Niveau eines mittleren Bildungsabschlusses entsprechen. Darüber hinaus enthält jede Unit mit den *Extra materials* auch anspruchsvolleren Lernstoff für Lernende mit fortgeschrittenen Englischkenntnissen. Ein Grundwortschatz von ca. 700 Wörtern wird vorausgesetzt. Dieser ist in einem gesonderten Vokabelverzeichnis aufgeführt.

**Mechatronics Matters** besteht aus acht Units, von denen jede eine in sich abgeschlossene Handlung enthält. Es ist möglich, einzelne Units auszulassen, jedoch wird empfohlen, wann immer Zeit und Lehrpläne es ermöglichen, die gegebene Reihenfolge einzuhalten, um einen kontinuierlichen Aufbau des Fachvokabulars zu gewährleisten.

In berufsbezogenen, praxisnahen Situationen werden die Lernenden mit allen wesentlichen Aspekten des Englischen vertraut gemacht, die ihnen in ihrem beruflichen Alltag begegnen können. Ob es um Kontakte zu Kunden und Geschäftspartnern, den Umgang mit technischen Dokumenten, das Schreiben von E-Mails, die Kommunikation über Maschinen, Messungen oder Motoren geht – **Mechatronics Matters** bietet in jedem Fall das passende sprachliche Rüstzeug.

Alle Units zeichnen sich durch kleine Lernschritte aus, so dass auch Lernende, die nur eine Wochenstunde Englischunterricht haben, ihre Kenntnisse und Fähigkeiten sorgfältig und zielstrebig aufbauen können.

**Mechatronics Matters** bietet die Möglichkeit eines handlungsorientierten und praxisbezogenen Unterrichts. Die aus dem Berufsleben gegriffenen, häufig authentischen Materialien vermitteln den Lernenden praxisnah das wichtige Fachvokabular. Die Lernenden schulen aktiv ihre Sprechfertigkeit in Rollenspielen und in Partnerarbeit, wobei neben der beruflichen auch die soziale Komponente des Arbeitsalltags einbezogen wird.

Grammatik wird in jeder Unit punktuell wiederholt, wobei die Regeln auf das Wesentliche begrenzt sind und insbesondere der Festigung von Grundstrukturen dienen.

Die Rubrik *Learning* bietet in jeder Unit Lerntechniken, die gezielt zeigen, wie man sich den neuen Wortschatz aneignet oder beim Verstehen und Schreiben sicherer werden kann. *Culture*-Kästen weisen auf interkulturelle Besonderheiten hin oder vermitteln landeskundliches Hintergrundwissen.

**Mechatronics Matters** vermittelt die englischsprachigen Grundpfeiler, auf denen sicheres Auftreten und korrektes Handeln im Berufsalltag basieren. Wenn die Lernenden sich dabei mit Vergnügen der englischen Sprache bedienen, hat dieses Werk sein Ziel erreicht.

1

# My company

## 1 Warm-up

**Work with a partner.**

**a What would you like to know when you meet a new colleague? Make a list and use it to get to know your partner.**

**b You start work at a new company and meet the following people. What are they doing? (The verbs in the box will help you.)**

soldering • checking • looking at • working with • testing • repairing

## 2 Introductions at work

2

**Nina Bauer, a mechatronics technician at Rinsieger GmbH, an engineering company, meets some new colleagues at the start of a visit to the company's subsidiary in Poland. Read or listen to the conversation.**

RECEPTIONIST Good morning. Can I help you?
NINA Good morning. My name's Nina Bauer. I have an appointment with Mr Sepczuk, the human resources manager.
RECEPTIONIST What time is your appointment?
NINA At half past eight.
RECEPTIONIST OK. You're a little early. I think Mr Sepczuk is still having breakfast in the canteen but I'll give him a call. Would you like to take a seat over there, please?
NINA Thank you.
...

MR SEPCZUK Ms Bauer? My name's Marek Sepczuk. I'm the HR manager. How do you do?

NINA How do you do, Mr Sepczuk?

MR SEPCZUK I hope you had a pleasant journey.

NINA Yes, thanks. I came by train. The trip was very interesting.

MR SEPCZUK Good. Now, can I introduce Anna Kozak? Anna is our training manager. She takes new employees through the induction programme.

ANNA Hello. I'm Anna. Pleased to meet you.

NINA Pleased to meet you, too.

ANNA Welcome to Poland.

NINA Thanks. It's nice to be here.

ANNA So, you usually work in Rüsselsheim, Nina.

NINA Yes, that's right. I'm based at headquarters.

ANNA Well, we're just a small subsidiary, but an important one.

NINA Oh, yes, of course.

ANNA Right. Let's go down to the shop floor and meet Lucjan Jakubiak. He's the production supervisor and he looks after the trainees. He's working in the mechatronics workshop today.

...

ANNA Hello, Lucjan.

LUCJAN Hi, Anna. Nice to see you. How are you?

ANNA Fine, thanks. And you?

LUCJAN I'm fine, too, thanks. Is this the new trainee from Germany?

ANNA That's right. Nina Bauer. Nina, this is Lucjan Jakubiak.

LUCJAN Good morning.

NINA Good morning, Mr Jakubiak.

LUCJAN Please, just call me Lucjan. You don't have to be formal with me.

NINA Thanks. I'm Nina.

LUCJAN OK, Nina. These are your colleagues, Magda and Tomek. A fourth trainee is joining us later. While we're waiting for him, I have a list of things to show you – the lathes, the production line robot, the CNC machine. Are you familiar with CNC machines? Do you know what we use them for?

## 3 Working with the text

**Are these statements correct or wrong? Correct the statements which are wrong.**

1 Nina is late for her appointment.
2 Mr Sepczuk is busy in his office when Nina arrives.
3 Mr Sepczuk is responsible for the staff.
4 Anna is a new employee.
5 The company's headquarters are in Poland.
6 The Polish branch of the company is quite small.
7 Lucjan Jakubiak is an apprentice in the workshop.
8 Magda and Tomek work in the office.

## 4 Working with words

**a Match the German words (1–10) with English words in the dialogue.**

1 Personalleiter/in
2 Ausbildungsleiter/in
3 Angestellte
4 Einführungsprogramm
5 Zentrale
6 Niederlassung
7 Produktionsstätte
8 Produktionsleiter/in
9 Werkstatt
10 Lehrling

**b Use the English words to complete this text.**

Nina usually works at her firm's ...[1] in Germany. At the moment, she's visiting a ...[2] in Poland. She's working on the ...[3]. Today she's in the ...[4]. Nina might become a ...[5] herself, like Lucjan, one day.

Anna is a ...[6]. She usually takes ...[7] through an ...[8]. She would like to become a ...[9] and look after ...[10].

## 5 Machinery in the factory

**a Lucjan talks about the machinery in the factory. Match seven of the words in the box with the photos on the opposite page. Which machine is not shown?**

CNC milling machine • conveyor belt • electrical discharge machine • laser welder • lathe • machining center • pulleys • robot

**b Now say which of the machines in the box Lucjan is talking about.**

1 'We use this to cut very hard metals.'
2 'This is a moving band used for transporting workpieces.'
3 'We join metals with this by melting the workpieces.'
4 'This one is used to shape metal parts. It's programmed by a computer.'
5 'Basically, it's just wheels with ropes. We use them for lifting heavy loads.'
6 'This is an automatic tool-changing facility – simply put, worktools on a rotating wheel.'
7 'This is our pick-and-place system. It's basically a mechanical arm.'
8 'We use this machine to spin pieces of metal which are then shaped.'

A

B

C
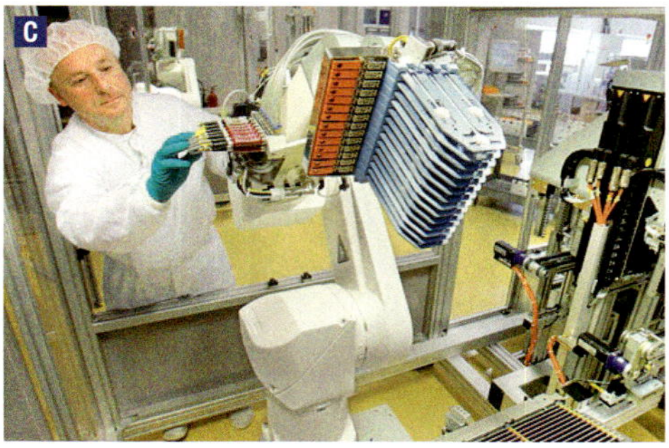

D

E
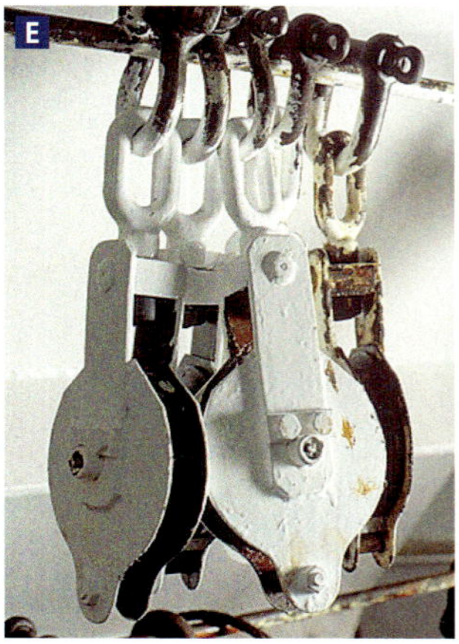

F

G

**c Describe other machines you know to your partner. Can he/she guess their names?**

Learning

## Wie kann ich mein Englisch verbessern?

Englisch kann man auf viele unterschiedliche Weisen lernen. Hier sind ein paar Tipps – aber bestimmt fallen Ihnen noch mehr Möglichkeiten ein!

- Filme in Originalfassung sind mit Untertiteln leichter zu verstehen – und vielleicht klappt es beim zweiten Mal schon ohne Untertitel!
- Ein Magazin für Ihr Hobby gibt es ganz bestimmt auch in englischer Sprache.
- Warum nicht mal eine SMS oder E-Mail auf Englisch schreiben?
- CDs haben meist Booklets mit Texten – wer mitliest, versteht mehr.

Language

## Making Introductions

**Greeting people**

| | | |
|---|---|---|
| *A:* How do you do? I'm ... | → | *B:* How do you do? My name is ... (very formal) |
| *A:* Pleased to meet you. | → | *B:* Pleased to meet you, too. (less formal) |
| *A:* Welcome to ... | → | *B:* Thank you. Nice to be here. |
| *A:* Nice to see you again. | → | *B:* Thank you. Nice to see you again, too. |
| *A:* How are you? | → | *B:* (I'm) fine, thanks. And you? |
| *A:* (I'm) fine, too, thanks. | ← | |

Good morning (6 a.m. – 12 noon)
Good afternoon (12 noon – 6 p.m.)
Good evening (6 p.m. – 10 p.m.)

**Introducing people**

Can I introduce (Anna Kozak)?
This is (Lucjan Jakubiak).
These are (your colleagues, Magda and Tomek).

## 6 Making introductions

2

**Put the words in the correct order to make sentences. Read or listen to the conversation on pages 6–7 again to check your answers.**

1 Nina morning my good Bauer name's
2 do how do you?
3 pleasant I journey had hope you a
4 introduce can I Anna Kozak?
5 meet you to pleased
6 Lucjan please me call just
7 colleagues are your these

## 7 Making small talk

**Match the responses (1–8) to the questions or statements (a–d). There are two answers for each question or statement.**

a How was your trip here?
b Is your hotel all right?
c I'm afraid the weather's not very nice here at the moment.
d I think you'll find the work here very exciting.

1 You're right. Is it always so cold here at this time of year?
2 Yes. I'm sure I will.
3 Yes, it's great, thanks. My room's very comfortable.
4 It was fine, thanks. I had a very pleasant flight.
5 It's not very good at home either at the moment.
6 Well, my train was delayed. But I got here in the end.
7 Yes. I'm really looking forward to getting started.
8 Yes, thanks. It's quite close to here and the staff are very friendly.

Culture

## Meeting and greeting

In Anglo-American business situations it's common to shake hands with colleagues when you first meet them. When you meet in private, people don't usually shake hands. It's always best to wait and see what your colleagues do, then do the same.

Use both your first name and your last name when you introduce yourself in English.
*Hello, my name's Florian Fuchs.*

## 8 Over to you

**Work with a partner. Practise introducing yourself and making small talk. When you have finished, change roles.**

| Partner A | Partner B |
|---|---|
| Begrüßen Sie B und stellen Sie sich vor. | |
| | Grüßen Sie zurück und nennen Sie Ihren Namen. |
| Sagen Sie, dass Sie erfreut sind, Ihr Gegenüber kennen zu lernen. | |
| | Erwidern Sie. |
| Fragen Sie, ob die Anreise angenehm war. | |
| | Sagen Sie, dass Sie eine gute Reise hatten. |
| Machen Sie etwas Smalltalk, z.B. eine Bemerkung zum Wetter. | |
| | Erwidern Sie. |
| Stellen Sie eine weitere Person vor: Mary Green. | |
| | Nennen Sie Ms Green Ihren Namen und begrüßen Sie sie höflich. Fragen Sie Ms Green, wo sie arbeitet und was sie macht. |

## 9 Company structure

3

**a Listen as Anna tells Nina about some people who work in Rinsieger's Polish subsidiary. Match the people (A–F) to their jobs (1–6).**

*Helena Konopka*

*Marek Sepczuk*

*Ewa Sawicka*

*Jakub Kolbe*

*Monika Olszewska*

*Lucjan Jakubiak*

**1** Production manager
**2** Dispatch supervisor
**3** Production supervisor
**4** HR manager
**5** Marketing manager
**6** Head of Accounts

**b** **Copy the organization chart below into your exercise book then listen again. Complete the chart with the names of the departments.**

- ...[1] Resources (HR) — Marek Sepczuk
  - Personnel
  - ...[2]
  - Training
- ...[3] — Helena Konopka
- ...[4] — Ewa Sawicka
  - Sales
  - Advertising
  - ...[5]
- ...[7] — Monika Olszewska
  - ...[8]
  - After-sales service
  - Packaging
- ...[6] — Jakub Kolbe
- ...[9] — Lucjan Jakubiak

RINSIEGER POLSKA Sp. z o.o.

## 10 Working relationships

**Use these words to complete the sentences.**

boss • closely • line • reports • responsibility • responsible

1 He is a mechatronics technician and he ... directly to the production manager.
2 The mechatronics production manager is ... for the work of the mechatronics team.
3 They are the production supervisors. They work ... with the production manager.
4 I work for him. He is my ... .
5 My ... manager is very strict.
6 She has overall ... for staff safety training.

## 11 Over to you

**a** **Draw a diagram showing the parts of your company that you know.**

**b** **Work in groups of four. Take turns to explain your diagram to the other students.**

**Useful questions**

- *Who is responsible for ...?*
- *Who is in charge of ...?*
- *Who heads / is the head of ...?*
- *Is your ... department also responsible for ...?*
- *How many people work in the ...?*
- *What does ... do?*

## 13 Health and safety at work

**Nina is studying the safety and warning signs used in the factory. Match the signs (a–j) to their meanings (1–10).**

*a b c d e f g h i j*

1 no smoking
2 hard hat area
3 first aid
4 fire extinguisher
5 danger: high voltage
6 emergency telephone
7 wear safety gloves
8 wear ear protection
9 wear goggles
10 wear safety boots

## 14 Health and safety guidelines

**a Complete these health and safety guidelines with the words in the box.**

access • area • attention • designated • distract • down • hazardous • instructed • loose • operate • spilled • serious

- Do not wear …[1] clothing or jewellery around machines. It may catch on moving equipment and cause a …[2] injury.
- Never …[3] the attention of another employee, as you might cause him or her to be injured. If it is necessary to get the …[4] of another employee, wait until it can be done safely.
- Keep your work …[5] clean.
- Shut …[6] machinery before cleaning or carrying out maintenance.
- Do not block …[7] to fire extinguishers.
- Do not …[8] machines or equipment until you have been properly …[9] and authorized to do so by your supervisor.
- Use …[10] passages when moving from one place to another; never take …[11] shortcuts.
- Clean up …[12] liquid, oil or grease immediately.

**b Explain the health and safety guidelines in German.**

**c Now look at the cartoons. Which of the health and safety guidelines mentioned above are being broken?**

# Extra material

## First aid – what to do in the event of an accident

Please read these instructions carefully so that you are able to give the correct assistance to a colleague who has been injured.

**Important:** In the event of serious head injuries or major body traumas, call an ambulance immediately.

**First aid boxes**

These are positioned and clearly marked in all areas of the factory. Please acquaint yourself with their contents and how they are used.

They contain:

- Sterile bandages
- Antiseptic ointment
- Eye pads
- Tape
- Sterile eye wash
- Safety pins
- Gauze pads

**Severe bleeding procedure**

Call an ambulance – do not try to clean the wound – apply firm pressure to the wound using a sterile bandage.

If bleeding persists, apply pressure to the major artery which supplies blood to the injured area.

**Burns procedure**

Minor burns: If the skin is unbroken, run cool water over the burn for a few minutes – cover the burn with a sterile bandage – seek professional medical assistance.

Major burns: Call an ambulance – in the event of the person not breathing, administer mouth-to-mouth resuscitation straight away – remove smouldering clothing – cover the burn with a cool, moist sterile bandage – do not place water, creams or ointments on the burn and do not break blisters.

**Electrical injuries procedure**

Minor shocks: Turn off the source of electricity – put the person in a comfortable lying position and keep them warm with a blanket – allow time for recovery.

Serious shocks: Call an ambulance – check pulse and breathing: if either has stopped or seems slow, administer cardiopulmonary resuscitation straight away – treat any burns as instructed above.

## 1 First aid box

**a** **Translate the first aid box items mentioned in the text into German. Use a dictionary to help you.**

**b** **Complete the sentences with the correct items from the first aid box.**

1 Use … to cover an eye after cleaning dust or other irritants from it.
2 Use … to treat a minor wound once you have cleaned it.
3 Use one or two … to fasten a bandage.
4 Use a piece of … to keep an eye pad or a bandage in place.
5 Use … to protect a wound once you have cleaned it and dressed it with a gauze pad.
6 Use … to clean an eye if it has been irritated by dust or similar irritants.
7 Use … to cover a minor wound. Keep them in place with tape or a bandage.

## 2 Life-saving treatment

**Find expressions in the text and match them to the diagrams.**

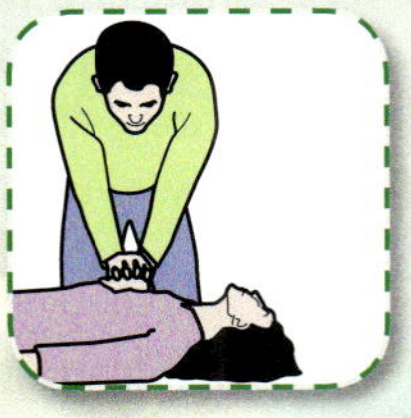

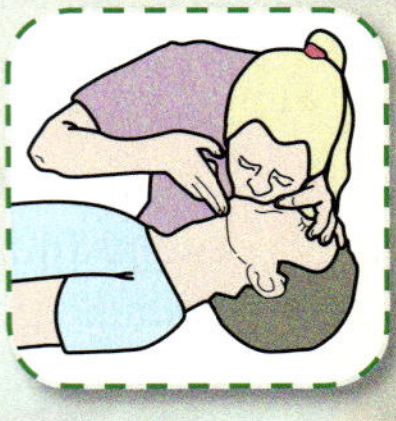

## 3 Mediation

**Work with a partner.**
**Take turns to explain the first aid instructions in German.**

## 4 Health and safety in your workplace

**Which dangers to health and safety do you face in your workplace? How does your company inform you about health and safety rules?**

2 

# In the workshop

## 1 Warm-up

**Work with a partner.**

**a** **Read the list of help and information you should get from your company when you start an apprenticeship. Add three more ideas of your own.**

**b** **Now decide which five items are the most important.**

**Help and information for new apprentices**

- Information about health and safety rules
- A welcome meeting with the human resources manager
- A tour of the factory
- Information about when you can take holidays
- A formal introduction to your colleagues
- Information about social activities organized by the company

## 2 A tour of the site

4

**Viktor Henschel, a mechatronics apprentice, arrives at Czech Chocolate SRO near Brno in the Czech Republic for three months' training abroad. Read or listen to the conversation with Damek Novak, a production engineer.**

DAMEK Hello, Viktor! Nice to meet you. My name's Damek Novak. Welcome to Brno!

VIKTOR Hello! Thank you. Nice to meet you too.

DAMEK Right. I'm going to take you to the mechatronics workshop in a few minutes. But before we do that we'll have a quick tour of the factory site to help you find your way around. Does that sound OK?

VIKTOR Sounds great!

DAMEK Good. We're in the administration block or 'admin' for short. This is where Sales, Purchasing, Marketing and Finance have their offices. That's the main reception desk on the left. Let's go up to the first floor – we can see the whole site from there. Follow me!

*They walk up the stairs to the first floor:*

DAMEK Now, do you see that big building directly opposite us?
VIKTOR Yes. Is that the production hall?
DAMEK Correct! Very good. We'll take a look at that in a minute. To the left of the production hall are the delivery bays, where our raw materials are brought into the factory.
VIKTOR Oh, yes. I see ...
DAMEK Can you see that row of tall buildings behind the production hall?
VIKTOR Yes.
DAMEK The building on the left, next to the delivery bays, is our Human Resources department or 'HR' for short. To the right of that is Research and Development or 'R&D'. Down at the far end on the right is our warehouse. Oh, and between that and R&D is the training centre.
VIKTOR It's a really big factory, isn't it? I can't wait to see the production hall!
DAMEK Well, that's where we're going now. We'll start with secondary production 1, which is between primary production on the left and secondary production 2 on the right. Let's go!

*Damek takes Viktor across to the production hall:*

DAMEK So, here we are in secondary production 1. This is where our products are processed.
VIKTOR Pretty impressive ...
DAMEK We work with full automation and we only use the most modern production-line robotics. You can see some of that in action here. But let's walk through to secondary production 2.

*They go into secondary production 2:*

DAMEK Right. Here we are in secondary production 2. As you can see, this is where our products are packaged, using the latest packaging technology.
VIKTOR Ah, yes. I see.
DAMEK And if we walk this way, we come to the loading bays where our chocolate is loaded onto lorries and taken to the shops.

*They leave the production hall through the loading bays:*

VIKTOR Thanks! That was really interesting. I'm going to learn a lot here, I think.
DAMEK Good. Here's the engineers' shop on the right, next to secondary production 2. The mechatronics workshop is on the ground floor. Let's go in and meet your new colleagues, shall we?
VIKTOR Yes, that would be great!

## 3 Working with the text

**Complete the sentences about the conversation in your own words.**

1 The sales department is ...
2 When a lorry arrives with raw materials, ...
3 In secondary production ...
4 You won't see many workers in secondary production 1 because ...
5 The packaging process is very efficient because ...
6 After the products have been packaged ...

## 4 The site plan

**Read the dialogue on pages 16–17 again and match six of the departments/areas in the box to numbers 1–6 on the plan. (You will use letters a–d in exercise 7.)**

changing rooms • delivery bays • engineers' shop • medical centre • research and development • secondary production 1 • training centre • warehouse

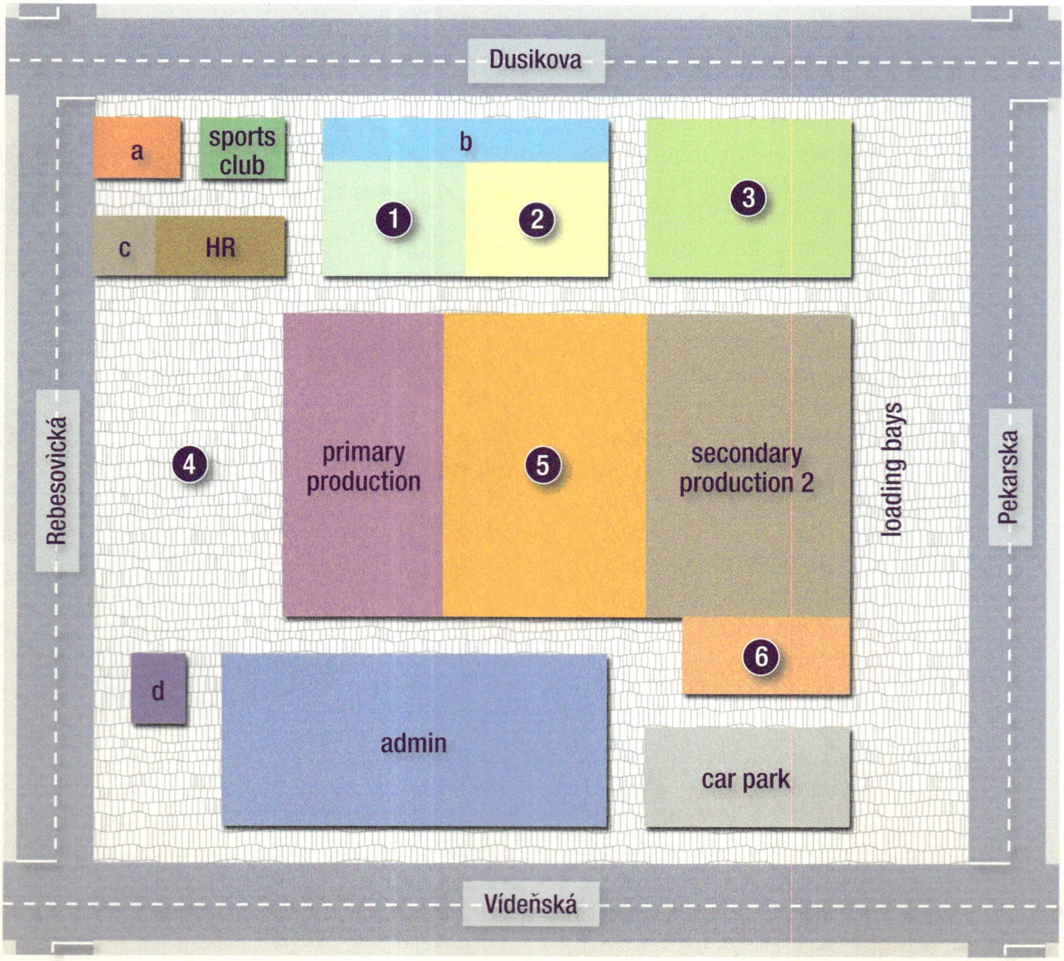

## 5 Working with words

**Match the expressions from the dialogue (1–10) with their meanings (a–j).**

1 administration block
2 reception desk
3 delivery bay
4 raw materials
5 research and development
6 warehouse
7 full automation
8 production line
9 ground floor
10 colleagues

**a** a sequence of machines and workers which together make a product
**b** the floor between the basement and the first floor of a building
**c** the department which develops new products and production techniques
**d** a system of production in which all the work is done by machines
**e** unprocessed substances which are used to make a product
**f** the part of a factory where management and commercial functions are performed
**g** the place where supplies are brought into a factory
**h** the building used for storing supplies or finished products
**i** the people who work together in an organization
**j** the point at which visitors to a company are welcomed

## 6 Where is it?

**Look at the site plan on page 18 and decide whether these statements are correct or wrong. Correct the wrong statements.**

1 The engineers' shop is on the same side of secondary production 2 as the warehouse.
2 Secondary production 1 is next to the primary production area.
3 R&D is opposite the warehouse.
4 The car park is beside HR.
5 The delivery bays are beside the training centre.
6 The training centre is opposite secondary production 1 and next door to R&D.

## 7 Finding the canteen

**Viktor has to find his way to the canteen. As you listen, follow the directions on the site plan (page 18). Decide which building (a–d) is the canteen.**

Language

### Describing locations

Culture

## Using first names

In English-speaking countries like the USA, the UK and Australia, colleagues normally call each other by their first names rather than their surnames. Trainees, however, should call a more senior member of staff by his/her surname, until they are told otherwise. Use *Mr* for a man and *Mrs* or *Ms* for a woman. *Ms* is used instead of *Mrs* when it is not clear whether the woman is married.

## 8 Giving directions

**Listen to Damek giving Viktor directions to the canteen again. Which expressions do you hear: a or b?**

1 **a** Can you tell me how to find the canteen?
  **b** Could you tell me how to get to the canteen?
2 **a** I'll take you there myself.
  **b** I'll tell you how to get there.
3 **a** Go out of the door here and turn left.
  **b** Go out of that door and turn right.
4 **a** Walk past the loading bays and then take the first on the left.
  **b** Walk round the loading bays and then take the second on the left.
5 **a** Go straight up and take the first on the right.
  **b** Go straight on, then take the second on the right.
6 **a** Then take the first left between HR and the sports club.
  **b** Then take the first on the left between the sports club and human resources.

## 9 Over to you

**Work with a partner and give each other directions.**
Partner A: Look at File 1 on page 88. Partner B: Look at File 7 on page 92.

## 10 The mechatronics workshop

**a An apprentice is being introduced to a mechatronics workshop. In which order do you think these things will happen?**

**New apprentices will:**

**a** watch a demonstration of workshop tools

**b** do a routine maintenance check

**c** go on a tour of the workshop

**d** read instructions for a routine maintenance check

**e** be introduced to their colleagues

**b Viktor returns to the workshop and meets Vavrin Dvorak, his new training supervisor. Listen as Vavrin tells Viktor what they are going to do and compare the order of events with your answers above.**

6 **c Listen again. Complete each sentence with the correct alternative(s).**

1 Jarek is ...
  a a production engineer.
  b going to be Viktor's work partner.
  c doing an apprenticeship.

2 If Viktor has any problems, ...
  a he should sort them out by himself.
  b Jarek will help him.
  c Vavrin will help him.

3 Today Viktor is going to learn about ...
  a the layout of the workshop.
  b the tools he'll need in his work.
  c health and safety rules.

4 Tomorrow morning, Viktor is going to ...
  a read about the maintenance check.
  b carry out the maintenance check with Jarek.
  c watch while Jarek carries out the maintenance check.

Grammar

## Talking about the future

1 I**'ll** just **go** and see if Jarek's here.
2 He**'s going to look after** you while you're here.
3 **Am** I **carrying out** the maintenance check myself?
4 You **won't have to** do it alone.
5 Jarek **is doing** the maintenance check with you.

- Das *will*-Futur steht für spontane Entscheidungen, Versprechungen und Voraussagen. Es besteht aus *will* + Grundform des Verbs. (1) Oft wird *will* zu *'ll* verkürzt. (1) Beachten Sie: *won't* = *will not*. (4)
- Das *going to*-Futur wird für Absichten oder Pläne verwendet. Es besteht aus *am/is/are going to* + Grundform. (2)
- Das *present continuous* wird für feste Verabredungen und Pläne benutzt. (3, 5)

## 11 Practice

**Decide which phrase in *italics* is more appropriate.**

1 I'm not sure how to start up this conveyor belt.
No problem – *I'll show / I'm going to show* you how to do it.
2 Has Viktor met the other mechatronics apprentices yet?
No. *He's meeting / He'll meet* them this afternoon at 3 o'clock.
3 Look! The warning light is flashing.
OK. *I'll shut down / I'm going to shut down* the machine and take a look.
4 Don't forget. It's time for the maintenance check on Production Line 1.
Don't worry. *I'll check / I'm going to check* the production line this morning.
5 What's the plan for this afternoon?
*We'll check / We're going to check* the production line.
6 What's happening about the problem on the packaging machine?
The service team *will repair / is going to repair* it tomorrow.
7 The new software has just arrived.
Excellent. *We're installing / We'll install* it right away.
8 Right. I've installed the new software. What next?
*I'll start / I'm starting up* the production line again and we can check if it's working.

## 12 What do they say?

**Listen again to the recording from exercise 10. Complete the sentences from the dialogue. Then identify the tense and say why it is used.**

1 I ... start by introducing you to Jarek.
2 If you have any problems, he ... help you to sort them out.
3 You ... spend a lot of time here in the next couple of months.
4 First thing tomorrow morning you ... carry out the check.
5 ... I carrying out the maintenance check myself?
6 You ... have to do it alone.
7 Jarek ... doing the maintenance check with you.
8 I ... just go and see if Jarek's here and then we ... get started.

## 13 Tools of the trade

**Jarek shows Viktor the tools he uses in his work. Match the English names for them with the numbers in the picture.**

Allen keys • folding rule • hacksaw • hammer • multimeter • needle nose pliers • Phillips screwdriver • power drill • pump pliers • round nose pliers • (slot-head / flat-bladed) screwdriver • side cutting pliers • soldering iron • spanner • vernier callipers • wire strippers

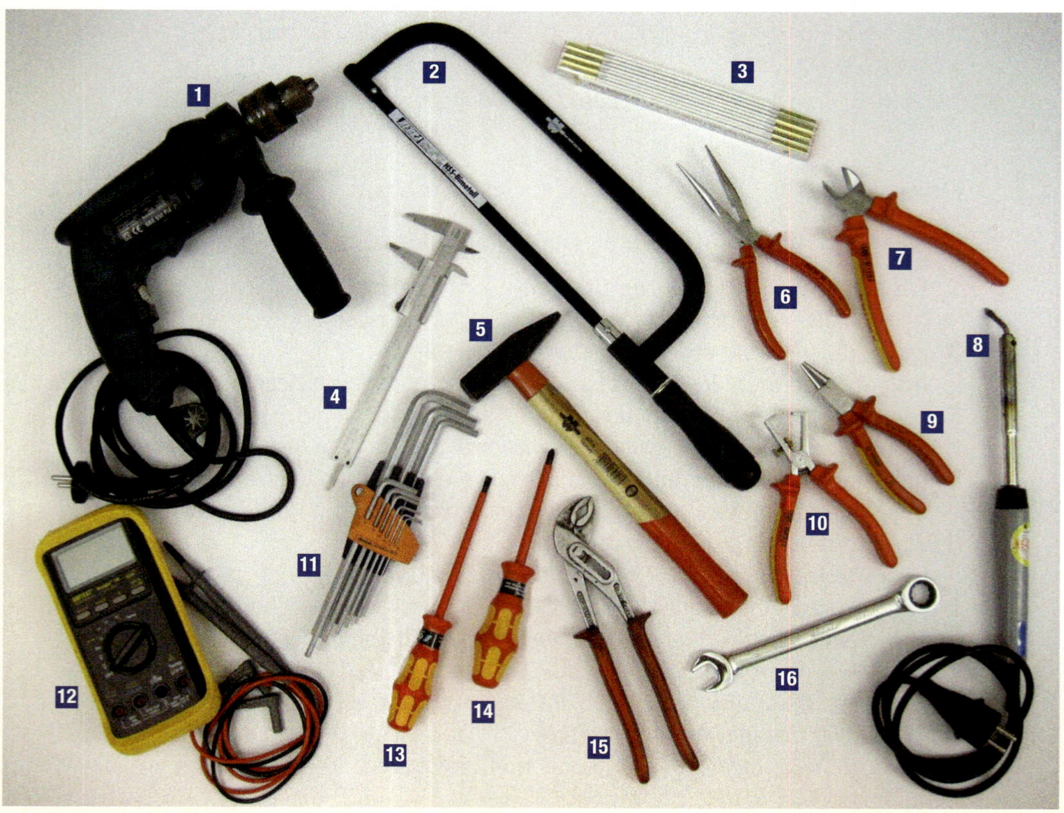

## 14 The right tool for the job

**Make sentences with the phrases in the table to say what the tools and instruments are used for. (Sometimes there is more than one possibility.)**

Example: *We use a screwdriver for tightening screws.*

| | | | | |
|---|---|---|---|---|
| We use | a screwdriver | for | drilling | components |
| | a multimeter | | gripping | electrical circuits |
| | a spanner | | loosening | holes |
| | a soldering iron | | measuring | large objects |
| | a pair of wire strippers | | picking up | nuts and bolts |
| | a power drill | | soldering | screws |
| | a pair of pliers | | stripping | small objects |
| | a folding rule | | testing | voltage |
| | an Allen key | | tightening | wires |
| | vernier callipers | | turning | |

## 15 Tool talk

**Discuss in a small group:**

- **Which of the tools on page 22 do you use and how often do you use them?**
- **What other tools and equipment do you use?**

Learning

### Wie man neue Vokabeln lernt

Eine sinnvolle Methode, ein neues Wortfeld (d.h. eine Gruppe von Wörtern aus einem bestimmten Themenbereich) zu lernen, besteht darin, sie um den jeweiligen Oberbegriff herumzuschreiben.

Oft prägen sich die Wörter durch die grafische Anordnung leichter ein, da der Zusammenhang im Vordergrund steht, nicht das isolierte Wort.

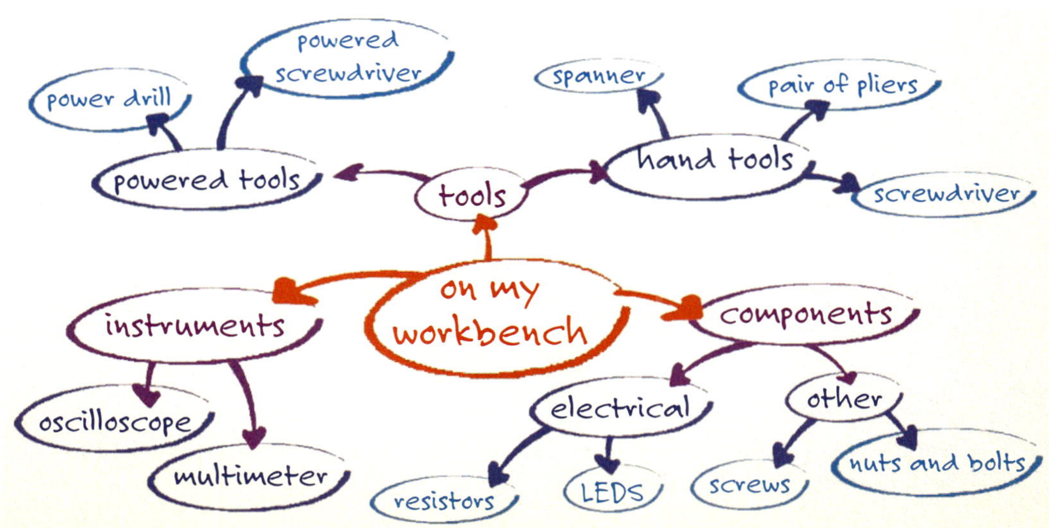

# 16 The maintenance check

Viktor is carrying out the maintenance check with Jarek as planned.

**a Match the symbols in the centre of the control panel to the maintenance tasks 1–6 in the table below.**

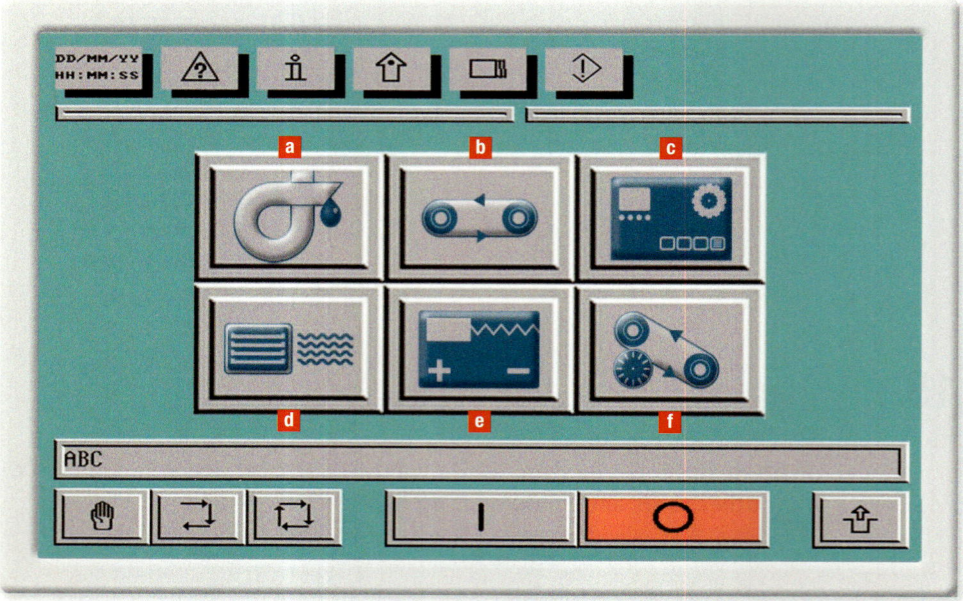


**b Use the words in the box to complete the maintenance table.**

blow • cloth • compressed • leakage • tension • waste

| No. | Position | Maintenance | Interval |
|---|---|---|---|
| 1 | All machines | Remove …[1], clean and remove surplus grease. | Weekly |
| 2 | Timing belts | Check …[2]. | Monthly |
| 3 | Conveyor belts | Clean with dry …[3]. | End of shift |
| 4 | Ventilator | Change or blow out the filter mats with …[4] air. | Monthly |
| 5 | Electronic cards | …[5] carefully with compressed (low pressure) air. | Half yearly |
| 6 | Water pipes and hoses | Check for water …[6]. | Weekly |

# 17 Mediation

**Work with a partner. Your partner does not understand the maintenance table. Explain it to him/her in German.**

Partner A: Explain 1, 2 and 3. Partner B: Explain 4, 5 and 6.

# Extra material

http://www.technical-apprenticeships.co.uk

Apple .Mac Amazon eBay Yahoo! News

## Technical training in the UK

### Becoming a mechatronics technician in the UK

The occupation of mechatronics technician is well established in Germany. Every year more than 5,000 young people take up apprenticeships to train for the job. German businesses have increasingly recognized the value of employing dedicated mechatronics technicians, a fact reflected in the growing number of jobs in the area.

However, the situation is a little different in the UK. There are far fewer jobs for mechatronics technicians: mechatronics work is often carried out by a combination of electricians and mechanics.

Duncan Blair is a college lecturer who teaches mechatronics in Scotland. 'The UK has an apprenticeship scheme which is basically similar to Germany's,' says Duncan. 'Unfortunately, though, there isn't an apprenticeship in mechatronics.'

The normal training route for mechatronics technicians in the UK is a Higher National Certificate (HNC) in mechatronics. You can study for an HNC at a College of Further Education, which is roughly equivalent to a German vocational college.

'The HNC takes two or three years, depending on whether you do it full or part-time. Some trainees are sent on the course by their employers. Others will find jobs once the course is over,' explains Duncan.

So what does Duncan see as the advantages of the British system of training mechatronics technicians?

'One of our real strengths is the depth and breadth we bring to students' training. Our people learn about all of the key mechatronics skill areas such as pneumatics, hydraulics, programmable logic controllers, computer-aided design, robotics, sensor technology and systems monitoring.

'In addition to this, we offer students training in communication skills and project management,' says Duncan, most of whose students go on to find jobs in Scottish manufacturing companies.

## 1 Working with words

**Find words in the text with the same meaning as these German words.**

1 Mechatroniker/in
2 Ausbildung, Lehre
3 Ausbildungsweg
4 Auszubildende/r
5 Arbeitgeber/in
6 Speicherprogrammierbare Steuerung
7 Sensortechnik
8 Fertigungsbetrieb

## 2 Advantages and disadvantages

**What do you think are the advantages and disadvantages of the German training system compared to the British system?**

## 3 Talking about your apprenticeship

**Work with a partner. You are describing the structure and content of your apprenticeship to a trainee from the UK. Your partner will listen and ask questions. When you have finished, swap roles.**

# 3 Measuring and testing

## 1 Warm-up

**a Match the names of these pieces of equipment to the pictures.**

**a** multimeter
**b** oscilloscope
**c** capacitance meter
**d** function generator
**e** wattmeter
**f** test light

1
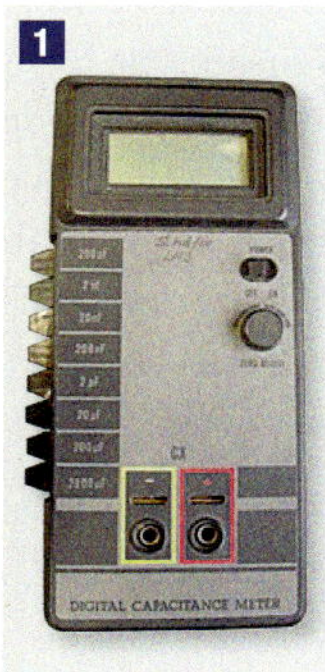


2

3

4
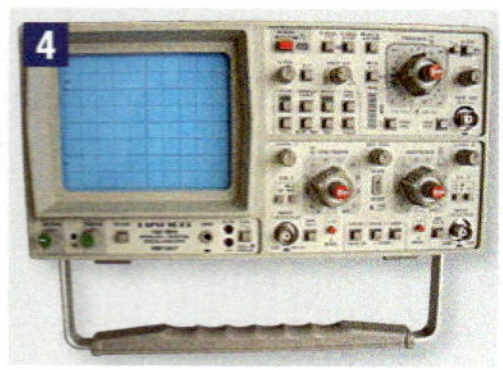

5

6
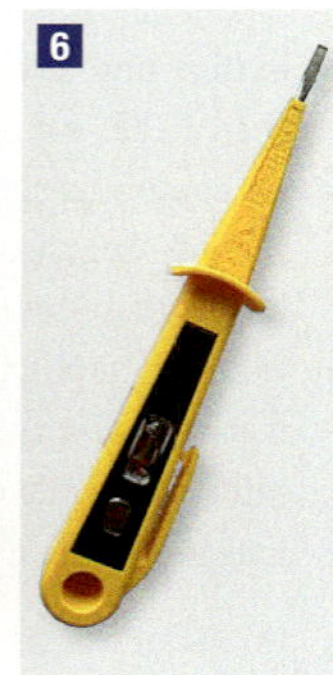

**b What can you use these things for? Make sentences.**

**1** You use a function generator
**2** You use a multimeter
**3** You use an oscilloscope
**4** You use a wattmeter
**5** You use a capacitance meter
**6** You use a test light

**a** to determine the presence or absence of electric voltage.
**b** to measure current, resistance or voltage.
**c** to generate electrical waveforms.
**d** to test newly-designed circuitry or to find out exactly which part of a piece of electronic equipment is faulty.
**e** to measure the electric power or the supply rate of electrical energy of any given circuit.
**f** to identify capacitors.

## 2 Buying a multimeter

7 **Tim Götz phones a British friend to ask his advice about a multimeter he would like to buy. Read or listen to the conversation.**

RECEPTIONIST Lincoln Electronics. Good morning. Can I help you?
TIM Good morning. Can I speak to Dave Mitchell, please?
RECEPTIONIST Who's calling, please?
TIM This is Tim Götz. I'm calling from Germany.
RECEPTIONIST Sorry. Could you spell your surname, please?
TIM Yes, of course. It's G O (with an umlaut) T Z.
RECEPTIONIST Thank you, Mr Götz. I'll put you through.

...

DAVE Dave Mitchell speaking.

TIM Hi, Dave. It's Tim here.

DAVE Hi, Tim. How are you?

TIM Fine, thanks. Listen. Do you have a minute? I'm thinking about buying a multimeter from a British internet firm. Could you help me?

DAVE No problem. Do you have the specs handy?

TIM Sorry, I don't understand.

DAVE The specs – specifications. You know, the details about the multimeter. I have to know what it's able to do before I can give you any advice.

TIM Oh, right. Hang on. I'll get them on the screen.

...

TIM Right. Are you still there? It's a digital multimeter. And it can measure the usual stuff – current, resistance, voltage.

DAVE OK. Anything else?

TIM Yes, it can also measure capacitance.

DAVE Sounds good. What's the Ohms range selection? Can it measure a broad range or only a narrow range?

TIM Well, it says, 200Ω to 2MΩ.

DAVE OK – a broad range. What other functions does it have?

TIM It says it has audible continuity, transistor and diode test facilities, overload protection and auto-ranging.

DAVE Has it got a data hold function?

TIM Yes, it has. You can check what you've measured before.

DAVE Good. That's useful. Now, it must conform to IEC-1010 standard. Does it?

TIM Yes, it does. How did you know that?

DAVE Well, it's working with overvoltage, so it has to conform to that standard. People aren't allowed to sell equipment if it doesn't conform to standard.

TIM Right. Good to know that. Is there anything else we should talk about?

DAVE Erm. What does it come with? Manual, probes, spare battery?

TIM Yes, there are probes with it and a manual and there's a holster, too, but there isn't a spare battery. It says here that the battery life is over 1,000 hours.

DAVE So you needn't worry about the battery at the moment. OK. What next? Weight?

TIM 0.75 kilogrammes.

DAVE Fine. How much is it?

TIM It's £32.

DAVE Sounds good, Tim. Go for it!

TIM Yes, I think I will. Thanks for your help, Dave.

DAVE No problem. Bye.

TIM Bye.

## 3 Working with the text

**a Finish the sentences using information from the text. Use your own words.**

1 Tim is phoning Dave because ...
2 Dave needs to know ...
3 Tim tells Dave that the multimeter can ...

**b Answer these questions on the text in your own words. Use complete sentences.**

1 Why is a data hold function useful?
2 Why must the multimeter conform to IEC-1010 standard?
3 What extras does the multimeter have?

## 4 Asking for information

**In pairs, role-play a phone call. You are going to help each other complete the specifications sheet for an oscilloscope.**
Partner A: Look at File 3 on page 89. Partner B: Look at File 9 on page 93.

Language

### Telephoning

**First contact**

I'm calling from Germany.
I'd like to speak to ..., please.
This is ... speaking.

**Connecting to another speaker**

One moment, please. I'll put you through.
Please hold the line.

**Explaining/Apologizing**

I'm sorry, ... is not available at the moment.

**Offering help**

Can I take a message?

**Promising action**

I'll get back to you about that.
I'll let Mr/Ms ... know.

**Dealing with difficulties**

Sorry, could you say that again, please?
Could you read that back to me, please?
Could you spell that for me, please?

**Closing the call**

All right! / OK! / Fine! / Right!
Thank you. / Thanks. Bye.

**TIP**

On the phone you can use the international telephone alphabet, but many English-speaking people simply use common words, such as *B for Butter*, when checking the spelling.

Spell email addresses slowly and carefully, letter for letter.

For example: t-j_pertrea@electro-star.co.uk =

T – **HYPHEN** – J – **UNDERSCORE** – P-E-R-T-R-E-A – **AT** – E-L-E-C-T-R-O – **HYPHEN** – S-T-A-R – **DOT** – CO – **DOT** – U-K

**TIP**

When people give you their telephone number, read it back to them to make sure you have written it down correctly.

When you give someone your telephone number, say each digit separately, except for double digits. For example: 01273 194488 ext. 10 = *Oh one two seven three, one nine double four double eight, extension one oh.*

Americans and Canadians say *zero* instead of *oh* for 0.

Learning

### Flüssiges Sprechen trainieren: *Shadow reading*

Hören Sie sich die folgende Aufnahme an und sprechen Sie mit (der Text steht auf S. 97–98). Dies kann auch als Rollenspiel in Partnerarbeit geschehen. Wiederholen Sie den Vorgang mehrere Male, während Sie die Aufnahme immer leiser stellen. Sprechen Sie den Text jedes Mal mit – nach ein paar Durchgängen sollte es ohne Mitlesen funktionieren. Wiederholen Sie den Vorgang so oft, bis die Aufnahme kaum noch zu hören ist und Sie den Text auswendig sprechen können. Die Methode des *Shadow reading* lässt sich mit allen Hörtexten aus dem Buch üben; bei längeren Texten reicht ein Ausschnitt.

## 5 Listening and speaking

8

**a** **Listen to the dialogue and follow the instructions in the Learning box above.**

**b** **What are the people talking about? Make notes.**

## 6 Role-play: Getting advice on the phone

**a** **What kind of problems might you have at work? Who would you ask for advice? Make notes for the role-play below.**

**b** **A mechatronics apprentice phones Chris Bennet, an experienced mechatronics technician, at Mech-Tech in London. Role-play the conversation in groups of three.**

| Partner A – Receptionist | Partner B – Caller | Partner C – Chris Bennet |
|---|---|---|
| Nennen Sie den Namen Ihrer Firma, begrüßen Sie B und fragen Sie, ob Sie helfen können. | | |
| | Nennen Sie Ihren Namen und sagen Sie, mit wem Sie sprechen möchten. | |
| Bitten Sie B, seinen/ihren Nachnamen zu buchstabieren. | | |
| | Erwidern Sie. | |
| Bedanken Sie sich und sagen Sie B, dass Sie ihn/sie durchstellen. | | |
| | | Nehmen Sie das Gespräch an und nennen Sie Ihren Namen. |
| | Sagen Sie, wer Sie sind. Fragen Sie Chris, ob er/sie Zeit für Sie hat. | |
| | | Bejahen Sie. |
| | Berichten Sie Chris von Ihrem Problem. | |
| | | Besprechen Sie gemeinsam mit B das Problem und erarbeiten Sie eine Lösung. |

Grammar

## Modal verbs

1 It **can** measure current, resistance and voltage.
2 **Could** you spell your surname, please?
3 It **must** conform to IEC-1010 standard.
4 You**'ll be able to** buy a new battery quite cheaply.
5 We **won't be allowed to** sell equipment that doesn't conform to standard.
6 We **should** talk about the problem.
7 Our instruments weren't up to standard so we **had to** buy new equipment.
8 This **mustn't** happen again.

Modale Hilfsverben haben folgende Eigenschaften:

- Sie haben in allen Personen dieselbe Form.
- Bei der Verneinung gibt es zusammengezogene Kurzformen. (5, 8)
- Alle außer *be able to* und *have to* werden mit dem Infinitiv ohne *to* verwendet.
- Wenn andere Zeitformen nötig sind, benutzt man die Ersatzverben *be able to* (für *can*), *be allowed to* (für *may*) und *have to* (für *must*). (4, 5, 7)

Achtung: *must not* = nicht dürfen, *doesn't/don't have to* = nicht müssen.

| Hilfsverb | Ersatzverb | Entsprechung |
|---|---|---|
| *can*<br>*could*<br>*may* | *be able to* | können<br>konnten/könnten<br>könnten |
| *may*<br>*might* | *be allowed to* | dürfen<br>dürfen |
| *must* | *have to* | müssen |
| *shall*<br>*should* | | sollen<br>sollten |
| *must not* | *not be allowed to* | nicht dürfen |
| *need*<br>*needn't* | | brauchen<br>nicht brauchen |

## 7 Practice

**Replace the German modal verbs in brackets with the most suitable modal verbs from the box above.**

1 It's very important to measure things properly. You ... (müssen) be 100% accurate.
2 You ... (nicht dürfen) leave equipment switched on overnight. It wastes energy.
3 This new piece of equipment is fantastic. It ... (können) measure almost anything.
4 You will ... (müssen) work faster. The boss is waiting for these results.
5 Irene ... (dürfen) try out the latest models at the exhibition last week.
6 We ... (nicht brauchen) give back the oscilloscope till tomorrow.
7 ... (Könnten) I use your multimeter? Mine's not working.
8 What ... (sollten) we do this afternoon? ... (Sollen) we tidy up the workshop?

Language

## Mathematical operations

| | |
|---|---|
| 1.1 | *one point one* |
| a > b | *a is greater (larger) than b.* |
| a < b | *a is smaller than b.* |
| $2 + 6 = 8$ | *Two plus six is (equals) eight.* |
| $9 - 3 = 6$ | *Nine minus three is (equals) six.* |
| $a \pm b$ | *a plus or minus b* |
| $3 \times 3 = 9$ | *Three times (multiplied by) three is (equals) nine.* |
| $16 \div 4 = 4$ | *Sixteen divided by four is (equals) four.* |
| $2^2 = 4$ | *Two squared is (equals) four.* |
| $2^3 = 8$ | *Two cubed is (equals) eight.* |
| $3^4 = 81$ | *Three to the power of four is (equals) eighty-one.* |
| $\sqrt{9} = 3$ | *The square root of nine is (equals) three.* |

- Note that in English commas and dots are used differently:

| | | |
|---|---|---|
| German | 4.000,00 Hz | 0,7V |
| English | 4,000.00 Hz | 0.7V |

- 0 is pronounced *nought* or *zero*.

## 8 Arithmetic

**a Work with a partner. Get from Start to Finish by moving through the boxes which result in 12. Which pair has finished first?**

| | | | | |
|---|---|---|---|---|
| START | | | | |
| six plus six | twenty-four divided by two | twenty minus nine | one plus ten | eight times three |
| seven times six | nine plus three | three multiplied by five | thirteen minus two | eighteen plus three |
| ten plus three | seventeen minus five | three times four | ten divided by two | ten plus four |
| eight times two | eleven plus two | forty-eight divided by four | thirty minus eighteen | twenty divided by five |
| one plus thirteen | nineteen minus five | ten times two | ten plus two | seven plus five |
| | | | | FINISH |

**b Now do the calculations below. First write them out in numbers.**

1 three squared plus seventy-seven
2 ninety-nine divided by three cubed
3 nine hundred and forty-one minus five hundred and fifty-five
4 the square root of twelve plus eight
5 nine cubed minus sixty-three
6 three to the power of four

## 9 Mathematical operations

**Work with a partner. You are going to read and write mathematical operations.**
Partner A: Look at File 2 on page 88. Partner B: Look at File 8 on page 92.

## 10 What is an Ohm?

**Work with a partner. Complete the definition with the words in the box.**

ampere • amperes • circuit (x2) • current (x3) • power • resistance (x4) • resistors • voltage • volts

**Ohm's Law**

Ohm's Law defines the relationship between (P) power, (E) voltage, (I) current and (R) resistance value through which one volt will maintain a current of one ...[1].

...[2] is what flows through a wire or conductor like water flowing down a river. ...[3] flows from points of low voltage on the surface of a conductor. ...[4] is measured in ...[5] or amps.

...[6] is the difference in electrical potential between two points in a ...[7]. It is the push or pressure flow through a ...[8] and is measured in ...[9].

...[10] determines how much current will flow through a component. ...[11] are used to control current levels. A very high ...[12] level allows a small amount of current to flow. A very low ...[13] level allows a large amount of current to flow. ...[14] is measured in ohms (Ω).

...[15] is the amount of current times the voltage level at a given point measured in wattage or watts.

*Adapted from: www.the12volt.com/ohm/ohmslaw.asp*

## 11 Equations and Ohm's Law

**Look at the pie chart and read the info box below. Match eight of the equations 1–12 in the pie chart to the equations a–h on the opposite page. Then write the four remaining equations out in words.**

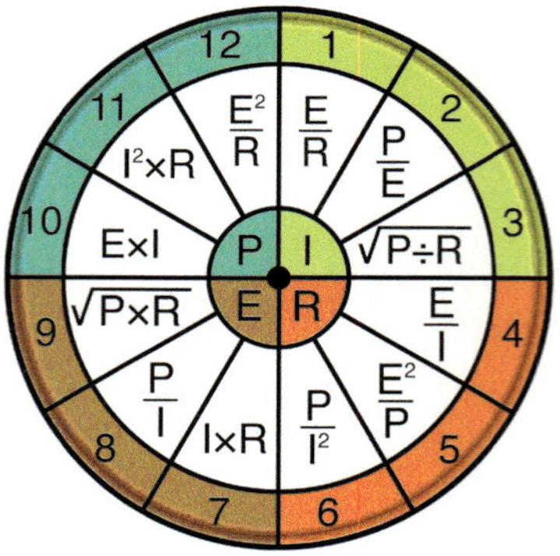


### Expressing Ohm's Law

Information

In exercise 10 you read about Ohm's Law. The pie chart above shows twelve different formulae in which Ohm's law can be expressed. Using it, you can define:

– (I) the **current** (segments 1–3), measured in **amperes**,

– (R) the **resistance** (segments 4–6), measured in **ohms**,

– (E) the **voltage** (segments 6–8), measured in **volts**, and

– (P) the **power** (segments 9–12), measured in **watts**.

For example, segment 1 shows the following equation: I = E/R.

**a** power = current squared *times* resistance
**b** power = voltage squared *divided by* resistance
**c** current = voltage *divided by* resistance
**d** current = the square root of power *divided by* resistance
**e** voltage = the square root of power *times* resistance
**f** voltage = power *divided by* current
**g** resistance = voltage squared *divided by* power
**h** resistance = power *divided by* current squared

## 12 Describing graphs

**Work with a partner.**

**a Use the table to describe the graphs. Sometimes you can use more than one expression.**

| | | | |
|---|---|---|---|
| The line | drops | drastically | at … |
| | flattens out | gradually | between … and … |
| | fluctuates | sharply | from … to … |
| | goes up/down | steadily | |
| | hits a maximum | suddenly | |
| | levels off | | |
| | peaks | | |
| | plunges | | |
| | remains constant | | |
| | rises | | |
| | shoots up | | |

**a**
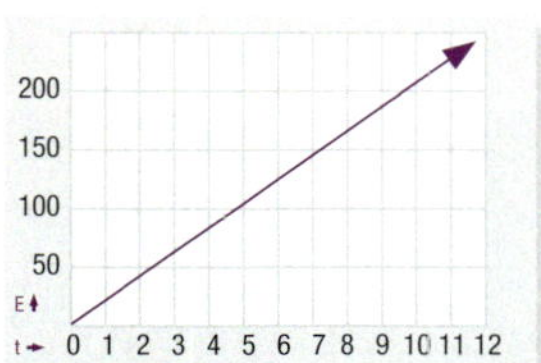


**b**
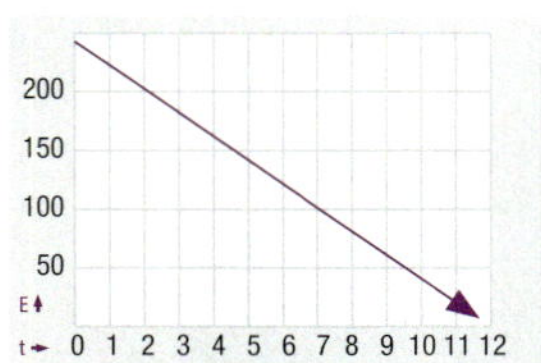


**c**
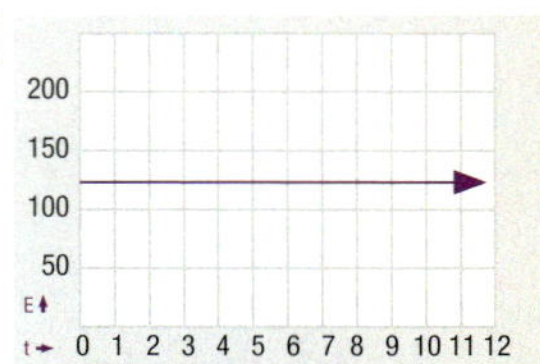


**d**
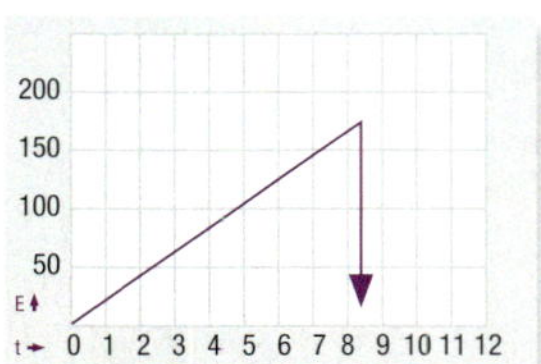


**e**
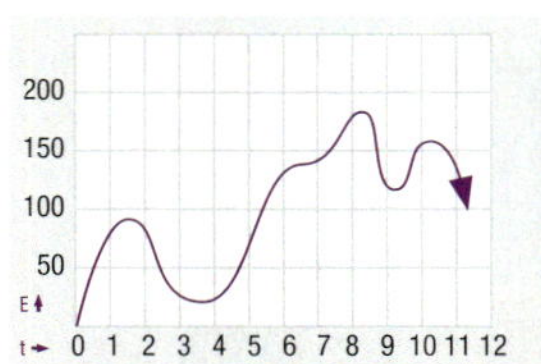


**f**
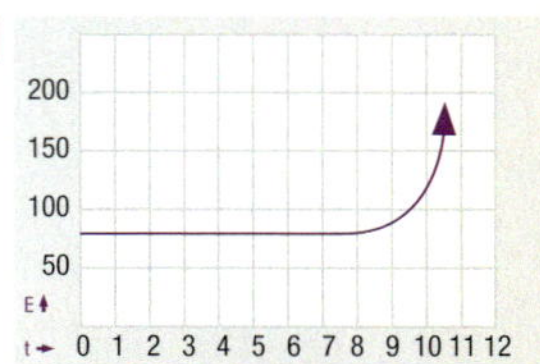


**g**
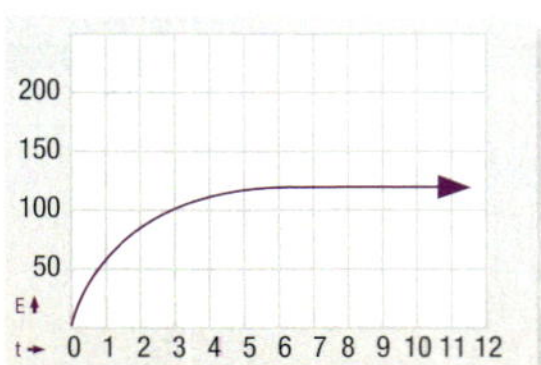


**h**
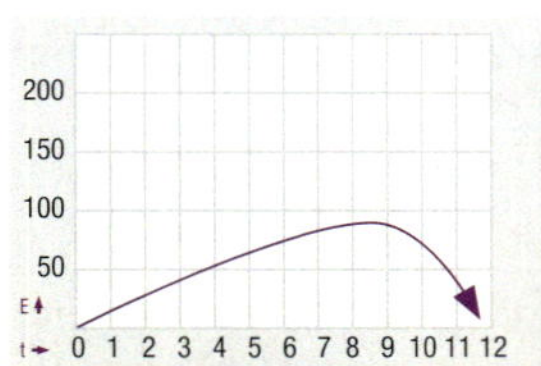


**i**
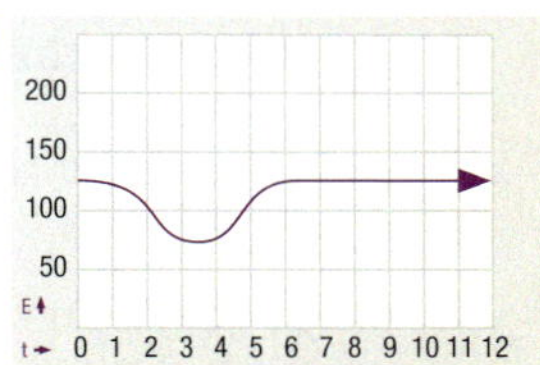


**b Now draw three graphs and describe them to your partner. (He/She mustn't see them!) Can your partner draw them correctly?**

## 13 Looking at the characteristic curve of a diode

**a Match the German words (1–5) to their English translations (a–e).**

1 Kennlinie
2 im Durchlassbereich
3 in Sperrrichtung
4 Durchlassspannung
5 Sperrspannung

a characteristic curve
b reversed-biased
c reverse voltage
d forward-biased
e forward voltage

9

**b Lukas and Magda are trying to find a fault in a diode. They are looking at a reading on an oscilloscope. Listen to them and say which graph they are talking about.**

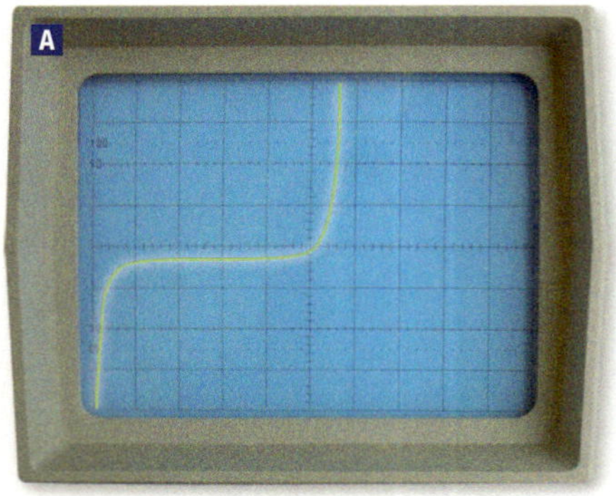


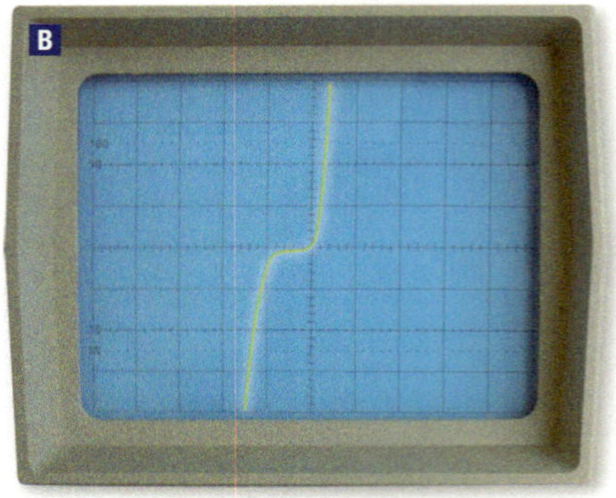


## 14 Comparing graphs

**a Which graph shows a characteristic thyristor curve? What does the other one show?**

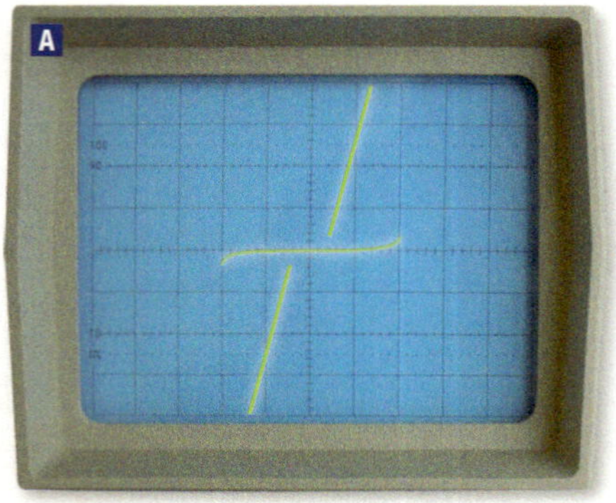


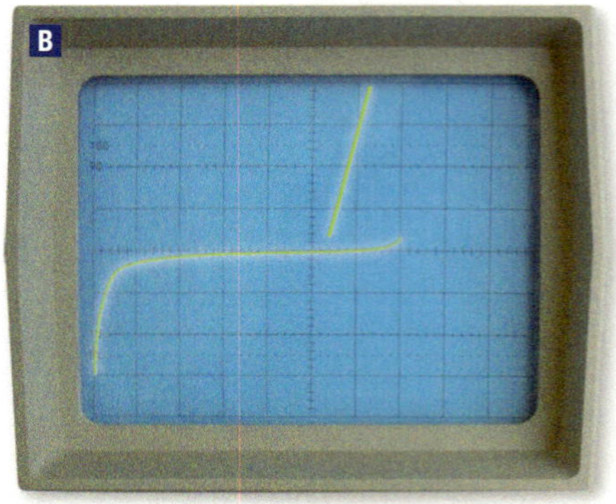


**b Describe the differences using these phrases and the phrases on page 33.**

| | |
|---|---|
| The curve on the left/right | swings/shoots/curves back/round.<br>doubles back on itself. |
| The top of the curve | mirrors the bottom of the curve.<br>is like a mirror image of the line below the zero/ neutral point. |

# Extra material

## 1 TPS

**a Where do you think the words in the box come from? English slang? Japanese business? Gangsta-rap? Klingon? Names of cars?**

heijunka • jidoka • kaizen • kanban • muda • poka-yoke

**b Read the text to find out. Then explain the meaning of the words in German.**

# Test the East!

### Japan leads the way in quality

A Japanese car? No thanks. Up until the middle of the last century, western car owners weren't keen to buy a vehicle produced in the East. By the beginning of the 1980s, however, 25% of cars bought in the USA carried the proud guarantee – 'Made in Japan'. Look around any street in any city today and you can't miss the Mitsubishis, Mazdas and Toyotas. Japanese cars are standing their ground against European and US vehicles.

What happened in the automobile manufacturing industry to change car buyers' minds about Japanese cars?

Think *kaizen* – continuous improvement. Along with *heijunka* (production smoothing) and *muda* (eliminating waste), the idea of *kaizen* was developed in the 1960s when the Japanese automobile firm Toyota started to work out a manufacturing and controlling philosophy aimed at producing top-quality cars which would sell for a reasonable price. The company eventually developed the idea of lean manufacturing and named this new way of working the Toyota Production System (TPS).

Here are some of the main ideas of lean manufacturing and TPS.

- *Jidoka* (zero quality control): human intelligence is programmed into automated machinery so that it is able to stop, start, load and unload automatically. The machines can also detect when a defective part has been produced, stop themselves and signal for help.
- *Kanban* (index cards): keeping record cards ensures that parts are only made in response to demand. Each workstation must have a card before it can start an operation.
- *Poka-yoke* (avoiding inadvertent errors): making sure that no defects or malfunctions occur during manufacture by, for example, eliminating choices that lead to incorrect actions; stopping a process if an error occurs; preventing damage to machines.
- *Kaizen* (continuous improvement): making sure that small, seemingly unimportant things that happen day after day in a business get better. Improving things on a daily basis will slowly lead to long-term improvements.

The philosophy of TPS can be adapted to fit any production environment. TPS experts suggest *jidoka* (achieving quality through the correct balance of human skill and intelligent automation) should always be part of a lean manufacturing system.

## 2 Discussion

**a Look back at the definition of *muda* (lines 15–16). What kind of 'waste' is produced in your company? How might the company improve things?**

**b What other things could be done in your company to improve production and other areas of work?**

# 4 Electrical components

## 1 Warm-up

**a Label the photos (1–5) with the words in the box. Then match the components to the definitions below (a–f). One definition does not match. What does it describe?**

resistor • transistor • diac • switch • thyristor

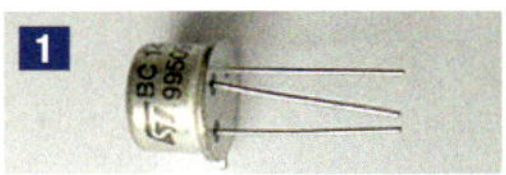

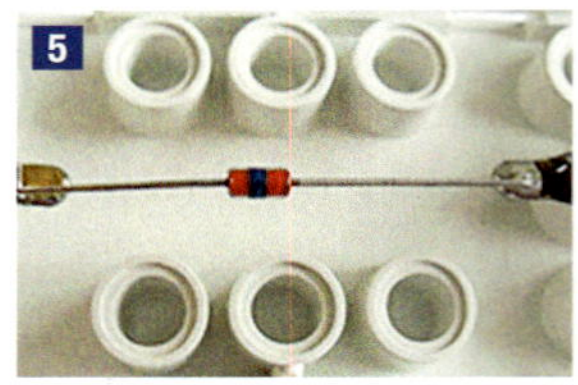

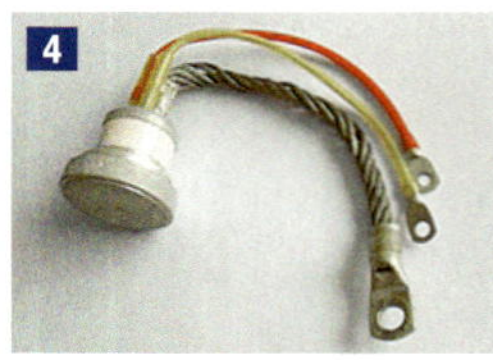


- **a** This component gives off light when a current flows through it.
- **b** Acts like a switch. It is made up of four layers of alternating N- and P-type material. They are mainly used where high currents and high voltages are involved.
- **c** They limit the flow of electric current in a circuit. Made from carbon or metal.
- **d** A small electrical component made of different layers of semi-conducting material, e.g. silicon. It is used as an electronic switch or amplifier. There are two types: npn and pnp.
- **e** This turns an electric current on or off.
- **f** This bi-directional diode is used to trigger a triac or an SCR. Typical power dissipations range from ½ to 1 watt.

**b With a partner, copy and complete this table. After your teacher has checked your tables, pin them on the wall. Which pair has the most components?**

| Name of component | Circuit symbol | What does it do? | Where can we use it? | Connect in series (S) or in parallel (P)? |
|---|---|---|---|---|
| resistor | | | | |

## 2 Solving a problem

Felix, a mechatronics technician, is trying to fix a problem with the arm of a robot.

**a Scan the text on the opposite page to find out which instrument from the box Felix uses to diagnose the fault.**

ammeter • mains tester • multimeter • ohmmeter • trimpot

Grammar

## The simple present and present continuous

**1** Anna **works** in Poland.
**2** I **buy** an English magazine once a week.
**3** **Are** you familiar with CNC machines?
**4** **Does** he **know** what we use them for?
**5** You **don't have to** be formal with me.
**6** Lucjan **is working** in the mechatronics workshop today.
**7** A fourth trainee **is joining** us later.

- Die einfache Gegenwart (*simple present*) wird für anhaltende Zustände (1) oder sich wiederholende Ereignisse verwendet. (1–5)
- Außer bei dem Verb *to be* (3) werden Fragen und Verneinungen mit dem Hilfsverb *do* bzw. in der 3. Person Singular mit der Form *does* gebildet. (4, 5)
- Die Verlaufsform der Gegenwart (*present continuous*) beschreibt Handlungen oder Vorgänge, die zeitgleich mit dem Sprechen/Schreiben ablaufen. (6)
  Das *present continuous* wird auch für feste Vereinbarungen und Pläne verwendet. (7)

**Signalwörter für das *simple present***
*always* (immer), *usually* (normalerweise), *normally* (normalerweise), *often* (oft), *sometimes* (manchmal), *hardly ever* (kaum), *never* (niemals)

**Signalwörter für das *present continuous***
*at the moment* (im Moment), *now* (jetzt), *just* (gerade), *this week/month* (diese Woche / diesen Monat)

## 12 Practice

**a Use the simple present or present continuous of the verbs in the brackets to make sentences.**

**1** Nina / a meeting with Lucjan this afternoon. (have)
**2** He / to the production manager once a week. (report)
**3** What's my job? I / robots. (design)
**4** I / in Russelsheim. (usually work)
**5** Lucjan / in the sales department. (not work)
**6** Tomek / as a mechatronics technician? (train)
**7** Mr and Mrs Sepczuk / in Poland? (live)
**8** Nina, Magda and Tomek / on the shop floor today? (work)

**b Now complete this description of one of Rinsieger's departments with the verbs from the box in the correct form. Then say which department it is.**

answer • file • handle • not be • organize • recruit

'In this department, we …[1] all the information on our personnel. We also …[2] new employees here. My colleague, Urszula Gajewska, …[3] the payment of salaries. She also …[4] employees' questions about pay, benefits and insurance. She's not here at the moment – she …[5] a training session for new employees. There …[6] many people in this department – currently, there are four, including myself.'

**b** **Now read or listen to the text. Which component(s) should be replaced?**

MR RICHARDS Ted Richards.

FELIX Hello, Mr Richards. It's Felix here. Do you have a minute? I'm calling from the main factory floor.

MR RICHARDS Oh, good morning, Felix. What can I do for you?

FELIX Well, I'm sorry to bother you but I'm having a bit of trouble diagnosing a fault here. It's the packaging robot. Can I ask you what you think?

MR RICHARDS Sure. Hang on, let me get the repair brief and the specifications for the robot.

...

MR RICHARDS Right. I have them. Ah, yes. I remember. We've had some trouble with the robot's left arm before. What have you done so far, Felix?

FELIX I checked the mechanics first. The arm isn't stuck in any way and there aren't any loose screws. Then I checked the electrics with the multimeter. They were all OK, except for a thyristor that was short-circuiting. I thought it was faulty, so I replaced it. The problem is, the fault's still there.

MR RICHARDS Have you checked the diacs?

FELIX No. I haven't.

MR RICHARDS Hm. Why don't you check them? It's equally important to check the diacs when it comes to fault-finding. I had a similar problem some months back. A diac had come loose from the circuit board and it wasn't sending out any trigger signals. Look closely at the circuit board, Felix. I'll bet that's what's causing the problem.

FELIX OK. Thanks for the tip. I'll check and get back to you.

...

FELIX Mr Richards? Hi, it's Felix here again. You were right, the soldering on one of the diacs was loose and it was hanging off the circuit board.

MR RICHARDS I thought so. OK. Well, check all the diacs in the robot you're working on to make sure they're all soldered properly. You'll have to check the circuit boards in the other robots very carefully at some point, too, just to make sure. Come up to my office after you've finished down there so that we can talk about when you can do that.

FELIX Right, Mr Richards. Will do. Thanks for all the good advice.

MR RICHARDS You're welcome, Felix. Glad to be of help. Bye.

## 3 Summarizing the job

**Felix has filled in his time sheet saying what he did today. In which order did he do the things? Put the notes (a–h) into the correct order with a partner.**

**a** Checked and changed a component.
**b** Checked diacs.
**c** Checked the moving parts.
**d** Phone call with Mr R. – advice needed.
**e** Planning session with Mr R.
**f** Re-soldered loose component.
**g** Tested the electrical parts.
**h** Tried to find out what was wrong with the packaging robot's left arm.

## 4 Working with the text

**Choose the most suitable ending to complete the sentences about the text.**

1 Felix calls his boss from
  **a** another factory.
  **b** the factory floor.
  **c** the workshop.

2 The problem with the arm of the robot is that
  **a** part of the arm is stuck.
  **b** there is a loose screw.
  **c** something is wrong with the electronics.

3 Felix has checked
  **a** the mechanical parts and all of the electronics.
  **b** the mechanical parts and some of the electronic components.
  **c** some of the electronic components.

4 He thinks that a thyristor is
  **a** loose.
  **b** causing the problem.
  **c** missing.

5 The problem with the diac is that it is
  **a** corroded.
  **b** old.
  **c** not connected to the other electronics.

6 As soon as he has repaired the arm of the robot, Felix has to
  **a** replace all the diacs in all the other robots in the factory.
  **b** check the other robots to see if there are any problems.
  **c** speak to Mr Richards about when to check the other robots.

## 5 Working with words

**a Find the English words and phrases which match the German words and phrases below. Check your answers in the text on page 37.**

1 einen Fehler diagnostizieren
2 Serviceunterlagen
3 lockere Schrauben
4 einen Kurzschluss verursachen
5 fehlerhaft
6 Platine
7 Triggersignale
8 Lötstellen

**b Now complete this text using the English words and phrases.**

A mechatronics technician often has to check a piece of electronic or mechanical equipment which is …[1]. First, the technician has to study the …[2] before he or she has a look at the problem.
It isn't always easy to …[3], but there are several things you should check. On the mechanical side, it makes sense to check for missing or …[4]. As far as the electrics are concerned, you should scrape the …[5] on the …[6] to see if it is still fixed.
Some components may be …[7]; test any diacs to check if they are sending out …[8].

Grammar

## Adjectives and adverbs

1 I'm calling from the **main** factory floor.
2 I thought it was **broken**.
3 We checked the mechanics **quickly**.
4 How **fast** can you do that?

- Adjektive beschreiben, wie jemand oder etwas ist, und beziehen sich auf ein Nomen. (1, 2)
- Adverbien bezeichnen, wie etwas gemacht wird, und beziehen sich auf ein Verb. (3, 4)
- Im Englischen werden die meisten Adverbien von einem Adjektiv abgeleitet, indem man *-ly* anhängt. (3)
- Das zu *good* gehörende Adverb ist *well*.
- Die Adverbien *hard*, *fast*, *early* und *late* sind identisch mit dem Adjektiv. (4)

## 6 Practice

**a Decide whether an adjective or an adverb is needed in the following sentences.**

1 Felix is a *smart/smartly* technician.
2 The workshop is very *quiet/quietly*.
3 Mr Richards found the fault *easy/easily*.
4 He did his training at a *famous/famously* college.
5 This is a *new/newly* robot.
6 Every component is working *proper/properly* now.
7 Some companies pay their employees very *bad/badly*.
8 I like to work in a *tidy/tidily* workshop.

**b Complete this text with *good* or *well*.**

Mr Richards is a ...[1] boss. He likes Felix because he always does a ...[2] job. He did very ...[3] in his exams and he gets on ...[4] with his colleagues. Everyone in Mr Richards' factory likes to do ...[5] work. They respect Mr Richards and know that he pays ...[6]. In other factories, conditions are not always so ...[7]. Everyone who works for Mr Richards knows that they'll be treated ...[8].

## 7 Industrial robots

**Complete this text about robots with six of the words in the box.**

assembly • motion • precise • precisely • programmer • segments • sequence

A typical robotic arm is made up of seven metal ...[1], joined by six joints. The arm moves very ...[2], repeating exactly the same movement over and over again. The robot uses ...[3] sensors to make sure it moves just the right amount. To teach a robot how to do its job, a ...[4] guides the arm through the motions using a handheld controller. The robot stores the exact ...[5] of movements in its memory, and repeats them again and again every time a new unit comes down the assembly line. Most industrial robots work in auto ...[6] lines, putting cars together.

Culture

## Being polite

English speakers use a lot of extra words to give a polite and helpful impression. This is particularly important on the telephone. If you give a short answer, your conversation partner may think that you are angry or rude.

- If you have to interrupt someone at work, ask if the other person has time for you.
- When making a request, try to start with, *Can/Could I* or *Would you mind ...* .
- Don't forget to use *please* and *thank you*.
- Try to avoid answering with just *yes* or *no* – always use three words, e.g. *Yes, I do. / No, we haven't.*

## 8 Role-play: Asking for help and advice on the telephone

**a Work with a partner. Read the situation and the dialogue and choose your role. Think about what you will say. Make notes, then role-play the telephone call.**

*Please check the DC motor in the JZ97. It is running in the wrong direction. Give me a call if you have any problems.*

| Partner A | Partner B |
| --- | --- |
| | Melden Sie sich am Telefon. |
| Grüßen Sie mit Ihrem Namen und sagen Sie, woher Sie anrufen. Fragen Sie, ob B Ihnen bei einem Problem weiterhelfen kann. | |
| | Grüßen Sie zurück und bejahen Sie. |
| Beschreiben Sie Ihr Problem mit dem Motor und bitten Sie B um Rat. | |
| | Erkundigen Sie sich nach dem Motortyp. |
| Erwidern Sie. | |
| | Besprechen Sie das Problem mit A. |

**b Now change roles and make another phone call based on this new situation.**

The automatic door isn't working. It may be the gears. Have a look at the sensors, too. If you can't find the fault, get back to me. I'll be in the workshop till 8 p.m. this evening.

Learning

## Notizen während einer Höraufgabe machen

Machen Sie sich während jeder Höraufgabe Notizen. So trainieren Sie z.B. auch während eines Telefonats in der Fremdsprache, wichtige Punkte festzuhalten, mit deren Hilfe Sie später einen mündlichen oder schriftlichen Bericht über das Gesagte geben oder darauf reagieren können (z.B. per Brief oder E-Mail). Folgende Punkte können wichtig sein:

- Name des Gesprächspartners / der Gesprächspartnerin
- Zeitpunkt
- Anlass für das Gespräch
- Was wurde besprochen?
- Einzelheiten wie Namen, Zahlen, Daten
- Was wurde entschieden?

Bevor Sie den Hörtext in Übung 9 hören, lesen Sie die Anweisung und machen Sie sich vorher eine Liste mit wichtigen Punkten, auf die Sie beim Hören achten müssen.

## 9 Who, where, when, why, what?

11

**a** **Felix is going to listen to a talk given by a representative of a company which makes electrical components. Before you listen, make notes about the information you expect to hear. Read the Learning box first.**

*Who is giving the talk?*

**b** **Tell another student in German what was said in the introduction to the talk.**

## 10 About semiconductors

**a** **Here are some of Mr Blum's notes (a–f). Read them and match them to the headings (1–3). There are two notes for each heading.**

| 1 silicon | 2 doping | 3 electron structure |
|---|---|---|
| | | |

*a carbon, silicon, germanium – each has four electrons in its outer orbital*
*b by mixing a small amount of an impurity into the silicon crystal*
*c these form perfect covalent bonds with neighbouring atoms, creating a lattice*
*d changes the behaviour of silicon and turns it into a conductor*
*e at the heart of any electronic device*
*f the main element in sand and quartz*

**b Complete this section of Mr Blum's notes with the words in the box.**

chips • computerized • doping • element • impurity • semiconductors • silicon • waves

You'll find ...[1] at the heart of microprocessor ...[2] as well as transistors. Anything that's ...[3] or uses radio ...[4] needs semiconductors. Most semiconductor chips and transistors are made with ...[5]. It's a very common ...[6]; it's the main element in sand and quartz. Silicon is not really a very good conductor but you can change its behaviour and turn it into a semiconductor by ...[7] it. This means that you mix a small amount of an ...[8] into the silicon crystal.

## 11 Which components are missing?

**Felix looks at a CCS catalogue. Which of the components on the catalogue page are not in the photo on the opposite page?**

Address: http://www.ccs-components.com/catalogue

CSS COMPONENTS

Home | Operators | Developers | Investors | Press | Corporate responsibility | Components | About

| | Type | | Price Euro. Cent | Order no. |
|---|---|---|---|---|
| **Diodes** | FKS 2 | | 1.20 | D 58 43 |
| | MKS 2 | | 1.26 | D 58 44 |
| | MKS 4 | | 1.41 | D 58 45 |
| | MKS 02 | | 1.65 | D 58 46 |
| | MKP 10 | | 1.72 | D 58 47 |
| **Electrolythic capacitors** | 20SV P22M | | 1.29 | EC 45 68 |
| | 16SV P330 | | 2.29 | EC 45 69 |
| | 2R5TP680M | | 2.49 | EC 45 70 |
| | 6TB300M | | 2.79 | EC 45 71 |
| | 10TB220M | | 2.99 | EC 45 72 |
| **LEDs** | 3 mm Ø | | -.07 | L 23 91 |
| | 5 mm Ø | | -.33 | L 23 92 |
| | 8 mm Ø | | -.72 | L 23 93 |
| | 10 mm Ø | | 1.15 | L 23 94 |
| | Duo-LED, 3 mm Ø/5 mm Ø | | -.72 | L 23 95 |
| **Transformers** | Primary | Secondary | | |
| | 2 × 115V | 2 × 6V/333mA | 8.45 | T 67 32 |
| | 2 × 115V | 2 × 9V/222 mA | 8.45 | T 67 33 |
| | 2 × 115V | 2 × 12V/166mA | 8.45 | T 67 34 |
| | 2 × 115V | 2 × 6V/500mA | 8.45 | T 67 35 |
| **Resistors** | ¼W 5% – 68 | | -.15 (25 pcs) | R 67 42 |
| | ¼W 5% – 100 | | -.25 (25 pcs) | R 67 43 |
| | ¼W 5% – 150 | | 1.00 (50 pcs) | R 67 44 |
| | ½W 5% – 100 | | -.75 (25 pcs) | R 68 45 |
| | ½W 5% – 820 | | -.50 (25 pcs) | R 68 46 |
| **Bridge rectifiers** | Type | | | |
| | KL 325 | | 2.84 | B 77 71 |
| | KL 375 | | 2.66 | B 78 26 |
| | KS 220 | | 3.35 | B 77 76 |
| | KS 325 | | 3.54 | B 74 42 |

Internet zone

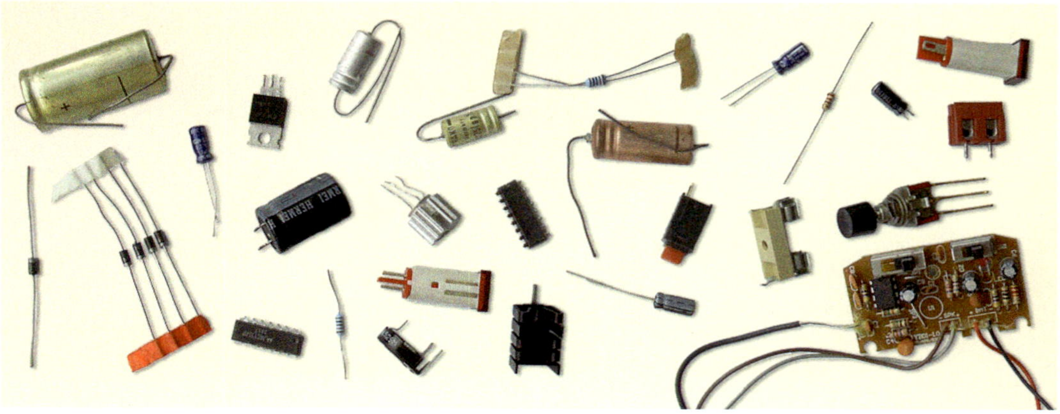

## 12 Circuit diagrams

**a What does a circuit diagram do? Choose the best definition.**

A circuit diagram shows

**a** how electronic components work.
**b** the values of all kinds of components.
**c** how electronic components are connected together.

**b Tell the class when, where and what you usually use circuit diagrams for.**

## 13 How to draw a circuit diagram

**a Complete these tips for drawing neat, clear diagrams with the words in the box.**

connecting • junction • lines • negative • positive • signals • supply • symbol • symbols • values

1. Make sure you use the correct ... for each component.
2. You may omit the battery or power supply ..., but you must include and label the supply ... at the top and bottom.
3. Draw ... wires as straight lines. Always use a ruler.
4. Put a blob (•) at each ... between the wires.
5. Label components such as resistors and capacitors with their ... .
6. The ... supply should be at the top and the ... supply at the bottom.
7. The negative ... is usually labelled 0V, zero volts.
8. If the circuit is complex, try to arrange the diagram so that ... flow from left to right: inputs and controls should be on the left, outputs on the right.

**b What is wrong with this circuit diagram? Use the tips above to describe the problems.**

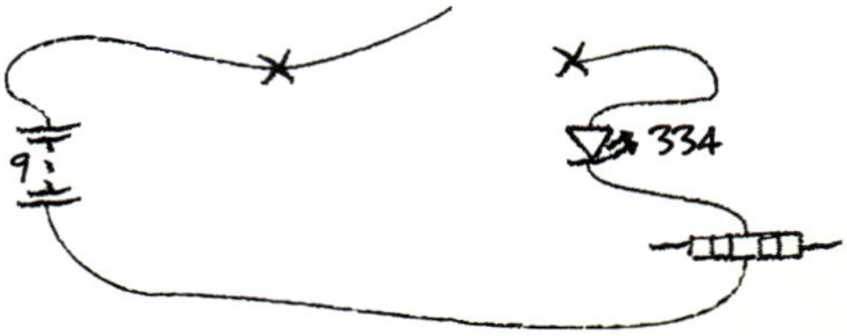

## 14 Identifying components

**With a partner, ask and answer questions about the components in the circuit diagram. Try to guess which component your partner is describing.**

Language

### Describing components

| | |
|---|---|
| What is it? | It's a … |
| What can it do? | It can conduct/insulate … |
| What is it used for? | It is used for conducting/insulating … |
| How is it connected? | It's attached/fixed/joined to … |

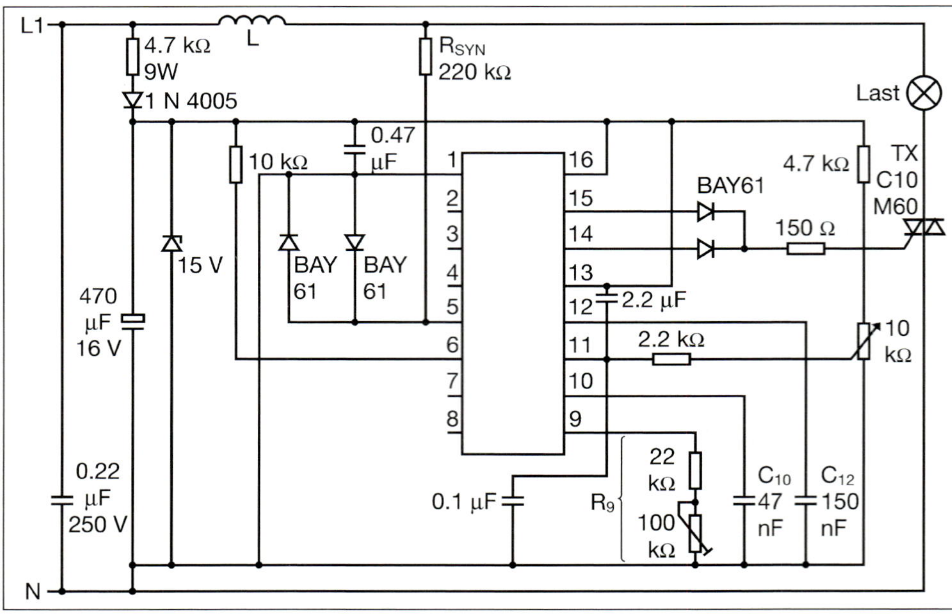


## 15 Describing a circuit diagram

**Work with a partner. Draw a circuit diagram and use the phrases below to describe it. Can your partner draw it without looking at it?**

Remember – inputs and controls are on the left and outputs are on the right.

The first element on the left-hand side is …

The … is in the middle / at the top right / at the bottom left.

Where does the … go?

Is there something here, in the middle?

From there, two connecting lines go off to …

Label the negative supply …

Is the … connected to the … ?

# Extra material

## The friendly face of robots

**a** DESCRIBE THE LAST ROBOT YOU WORKED ON. What was its function? What did it look like? Most mechatronics technicians have to install or maintain 'pick and place' robots in robotic cells on the factory floor. In the near future, however, you may be called to someone's home to work on a 'servant' robot.

**b** It sounds like a great idea – a robotic servant which keeps the house clean, entertains us and looks after the children. But before they become part of our everyday life, engineers have to design robots which humans feel comfortable with. As one roboticist puts it, 'It's not enough that the robot does different things in the house, that robot should also be able to perform these activities in a way that is comfortable and acceptable to people.'

**c** To find out just what is acceptable to humans, engineers have set up an experiment called The Robot House. Volunteers spend time there to see how they get on with the resident robot. One of the questions which the engineers have asked the volunteers to think about is what the home robot should look like. 'A little bit human with a face containing a mouth and eyes that light up to give a kind of expression,' is the answer. The volunteers found it less easy to relate to a purely mechanical shape.

**d** At a recent conference on robotics, one engineer said, 'The real challenge is to come up with robots that can actually communicate with people. To do that we have to solve an enormous number of technical problems – everything from understanding speech and gestures to making eye contact and having an awareness of a human's personal space,' he said. However, giving a robot too much intelligence of any sort is a frightening thought for most people.

1

2

3

## 1 Working with the text

**a** **Match the photos (1–3) with suitable paragraphs in the text.**

**b** **Describe the idea of robotic servants in your own words.**

## 2 Robotic servants

**In groups, decide what you would like a robotic servant to do for you. Then make a list of the kind of characteristics you would like your robotic servant to have.**

# 5 Installation

## 1 Warm-up

**a Look at the photo on the opposite page. What do you think this machine does? (Tip: you can find the answer in the dialogue below.)**

1 lift heavy equipment or boxes
2 X-ray people
3 produce tubes, e.g. for toothpaste
4 lay fibre optic cables

**b Now identify parts 1–8 of the machine using the words in the box.**

> compressed-air pistol • compression moulding unit • elevator • mandrels • operating panel • tube body maker • tube heading machine • waste suction unit

**c Put the actions which the machine performs into the correct order.**

a The tubes are held firmly, ready for the shoulders to be fitted.
b Waste is sucked into the waste suction unit.
c An elevator takes the caps up to the top of the machine.
d The machine makes the body of the tubes.
e The caps are loaded into a bowl, then led down to the tubes.
f The tubes are transported to the heading machine.
g A compression moulding system welds the shoulders on to the bodies of the tubes.

## 2 Arranging an installation

12

Julius Keller from PackSys Global (Switzerland) Ltd phones Alain Dufour at Orbis in Colmar to arrange delivery and installation of a new machine.

**Read or listen to the dialogue.**

JULIUS Hello, Alain. Julius from PackSys here.
ALAIN Hi, Julius! Nice to hear from you again. How are you?
JULIUS Fine thanks. And you?
ALAIN Very well, thanks.
JULIUS Glad to hear it. Alain, I'm just calling to confirm the arrangements for delivering, installing and commissioning the tube production system.
ALAIN Ah, yes. I got your email on Friday. That all looks pretty clear.
JULIUS Good. Now, as I said in my mail, our team will arrive on the evening of the 14th June. They'll take a look at the site for the machine that evening and spend the 15th making all the necessary preparations. They'll carry out the installation the next day.
ALAIN Great. One thing, though – our contractor is coming in two days later than planned to prepare the foundations on the factory floor. Will there still be enough time for the foundations to dry out before your team starts work?
JULIUS Ah, let me see ... they were originally coming on the 9th, weren't they? Hmm,

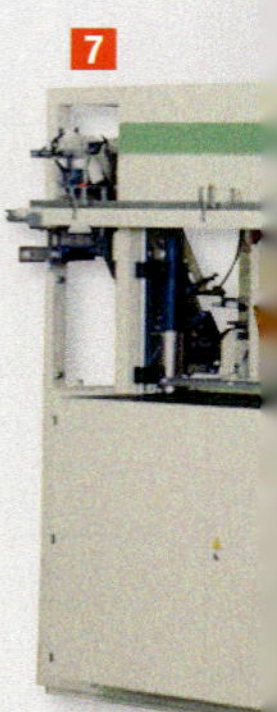

two days later ... the 11th. Yes, I think that's OK. If the foundations have three days to dry out, we'll be able to carry out the installation as planned.

ALAIN Do you still think the installation and commissioning is doable in one day?

JULIUS Oh, yes. I think so.

ALAIN I'm asking again because we would like the machine to be fully operational by the 17th. We have a heavy production schedule coming up and we might have some problems if there was a delay in getting the machine up and running.

JULIUS Don't worry, Alain. We'll do our absolute best to ensure that all systems are go by the 17th.

ALAIN Thanks, Julius. So, was that it?

JULIUS There is one more thing. Do you have the equipment for moving the machine?

ALAIN Yes, we've got it all lined up. If I understood you correctly, we'll need a forklift to transport the machine through the factory, but we don't need to worry about a crane for lifting the machine off the lorry, is that right?

JULIUS That's right. The lorry has a crane on board, so it'll just be a case of lowering the machine over the side onto the forklift.

ALAIN Excellent. Well it seems that we're all ready to go then, doesn't it?

JULIUS Yes, it does. If you have any questions between now and the 14th, you should get in touch, of course.

ALAIN Will do, Julius. Thanks.

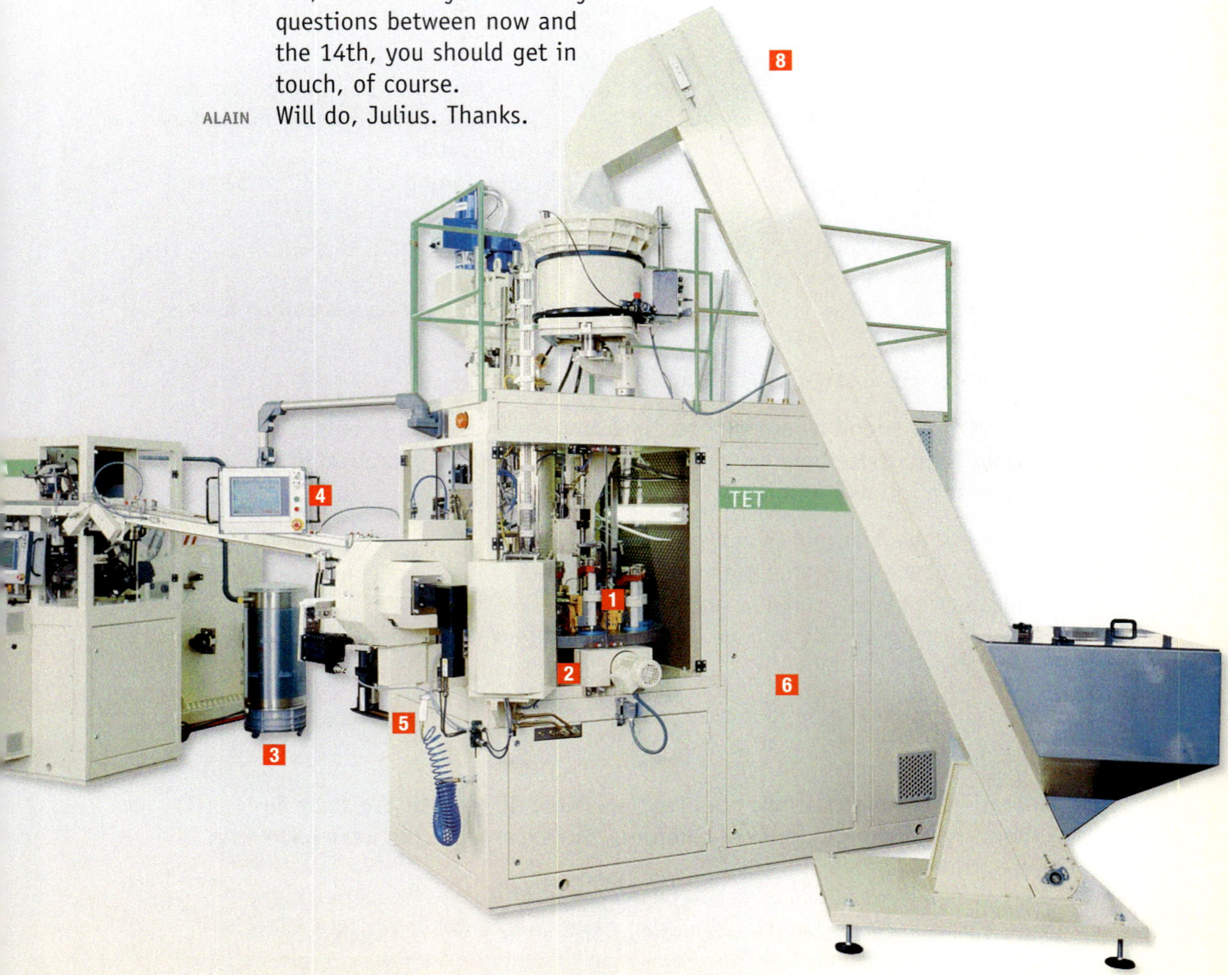

## 3 Working with the text

**Are these statements correct or wrong? Correct the wrong statements.**

1. Julius is calling to change the arrangements for installing the machine.
2. Last Friday Alain wrote Julius an email asking him about the machine.
3. The team from PackSys will carry out the installation on the 16th.
4. The contractors will lay the foundations on the factory floor on the 9th.
5. The foundations should be laid three days before the machine is installed.
6. If the machine is not working by the 15th, Orbis will have problems meeting their production schedule.
7. Julius thinks that there may be some problems in commissioning the machine by the date agreed.
8. Alain should arrange for a crane to be at the side of the lorry to lower the machine onto the forklift.

## 4 Working with words

**Find words and expressions in the dialogue with the same meaning as the definitions below.**

1. to start something up (e.g. a new piece of machinery) and ensure that it is working
2. to do something (such as an installation)
3. a company which does work or which provides a service for another company
4. the base of a building or the surface under a piece of heavy machinery
5. achievable
6. functioning and ready to be used
7. timetable
8. working
9. machinery
10. available to be used when needed

## 5 Writing an email

**An American friend would like to know how your company deals with installations. Write your friend an email describing an installation job. Include the following points:**

- types of machines or equipment installed
- where/how the jobs are planned
- installation jobs you know about
- installations which you have worked on and how you were involved

## 6 The EU's safety guidelines for installing industrial machinery

**Isabell Frey, a mechatronics apprentice who will be with the team in Colmar, reads the guidelines. Complete the excerpt on the next page with these words.**

avoid • limit • preventable • refitting • relating • risk • source • subject

### 1.5.1. Electricity supply

Where machinery has an electricity supply it must be designed, constructed and equipped so that all electrical hazards are …[1].
The specific rules in force …[2] to electrical equipment designed for use within certain voltage limits must apply to machinery which is …[3] to those limits.

### 1.5.2. Static electricity

Machinery must be so designed and constructed as to prevent or …[4] the build-up of potentially dangerous electrostatic charges and/or be fitted with a discharging system.

### 1.5.3. Errors of fitting

Errors likely to be made when fitting or …[5] certain parts which could be a source of …[6] must be rendered impossible by the design of such parts or, failing this, by information given on the parts themselves and/or the housings. The same information must be given on moving parts and/or their housings where the direction of movement must be known to …[7] a risk. Any further information that may be necessary must be given in the instructions.
Where a faulty connection can be the …[8] of risk, incorrect fluid connections, including electrical conductors, must be excluded during design and development or by giving information on the pipes, cables, etc. and/or connector blocks.

## 7 Two-word phrases

**Make two-word phrases from the two columns. Then check them in the text.**

| | |
|---|---|
| electrical | charge |
| electricity | hazard |
| electrostatic | limit |
| moving | parts |
| voltage | supply |

## 8 Mediation

**Summarize the text above in German.**

Learning

### Was versteht man unter Mediation?

Manchmal werden Sie aufgefordert, den Inhalt eines Textes im Deutschen wiederzugeben. Dann wiederum gibt es Dialoge, in denen Sie auf Deutsch formulierte Redeanlässe („Sagen Sie, dass Sie …") im mündlichen Gespräch auf Englisch ausdrücken sollen. Dieses sinngemäße Übertragen von einer Sprache in die andere nennt man „Mediation", also Vermittlung zwischen zwei Personen, die der Sprache des anderen nicht mächtig sind.

Wichtig bei einer Mediation ist die Konzentration auf das Wesentliche – stilistische Feinheiten sind hier zweitrangig. Übersetzen Sie nicht Wort für Wort – es geht darum, den Sinngehalt des Originals vollständig und sprachlich angemessen auf Deutsch wiederzugeben. Wenn Sie etwas Vorbereitungszeit haben, sollten Sie unbekannte Wörter im Wörterbuch nachschlagen.

## 9 Action planning

**Work with a partner. Julius is preparing an action plan which will assist him with the installation project at Orbis.**

**a Does column A give details about**

- extra points to be taken into consideration?
- things that have to be done for the project to be successfully completed?

**b Copy the plan into your exercise book. Then complete the missing parts (1–9) with the words in the box.**

book • brief • check with Orbis • confirm • ensure conformity • organize Zentis Logistik • phone • remind Orbis • tell

| Action plan – JK – 15/05 | |
|---|---|
| **A Action** | **B Notes** |
| 1 … production department that we need a tube production system ready for delivery to customer by 16/06 | h |
| 2 … the hotel to arrange accommodation | |
| 3 … installation dates with customer | |
| 4 … to have forklift on hand for installation date | |
| 5 … that foundations will be laid in time | |
| 6 … to provide overland transport | |
| 7 … with EU installation safety guidelines | |
| 8 … installation team | |
| 9 … flights | |

## 10 Notes on the action plan

**Work with a partner. Complete column B with Julius's notes (a–i). Each note matches one of the actions on his plan. The first one has been done for you.**

**a** Orbis has forklifts on the factory floor.
**b** Check before briefing team. / Order copies of guidelines for team.
**c** Make sure that Orbis gets this done at least five days before the installation. (Drying out time!)
**d** Don't forget! ZL must be booked at least ten days before a delivery.
**e** Important! But has to wait until everyone is back from holiday.
**f** Remind G. to book hotel with swimming pool.
**g** Remind G. that JL flying on to Paris after commissioning machine.
**h** Must be done immediately! Production needs one calendar month.
**i** Important! Orbis must confirm all their installation dates.

## 11 Safety notes for transporting, installing and commissioning the machine

**Work with a partner and match the safety notes (a–r) to their appropriate headings (1–4).**

1 Transportation and lifting

3 Mounting and checking safety devices

2 Installation

4 Commissioning / Trial run

- **a** Connect electrical cables according to the circuit diagram.
- **b** Ensure the machine does not get heavy jerks during the handling. The minimum capacity of the crane/forklift must not be less than 5,000 kg.
- **c** All installation work must be carried out by authorized personnel. Current accident prevention regulations must be observed.
- **d** The means of transport must withstand a load of at least 5,000 kg. When securing the lifting device, the machine's centre of gravity must be taken into consideration.
- **e** Before starting up operation, ensure that there is no one in the operating or danger area.
- **f** The machine should be lifted carefully by means of a crane/forklift.
- **g** Take care when lifting the equipment. No one should be under the lifted equipment. If the lifted equipment is brought down or it falls, it can endanger the life of anyone underneath.
- **h** The machine should only be connected to the electrical supply by skilled personnel.
- **i** Check direction of motor rotation by briefly pressing the motor switch.
- **j** Connect inlet and return lines of the cooling water to the works-side of the cooling system. Do not mix up the connections.
- **k** Connect the pneumatics lines to the works-side compressed air supply. Lead the pneumatic lines through as shown on the pneumatic plan.
- **l** All safety devices must be fully mounted before starting operation. The type and position of safety devices are described in the assembly instructions.
- **m** Mount and connect the Emergency Stop switch in the assigned position.
- **n** Avoid damaging the machine components during lifting. If necessary use pads to protect the machine components.
- **o** Before starting-up, all safety devices and safety measures necessary for machine operation must be checked for effectiveness by authorized personnel.
- **p** Set relief valves according to the hydraulic and pneumatic parts.
- **q** Switch on using main switch "ON".
- **r** Use wire ropes with appropriate capacity and length to avoid overloading the ropes.

## 12 What is going wrong?

**Look at the pictures. Say in your own words what is happening. Then check the safety regulations on page 51 for the correct procedures.**

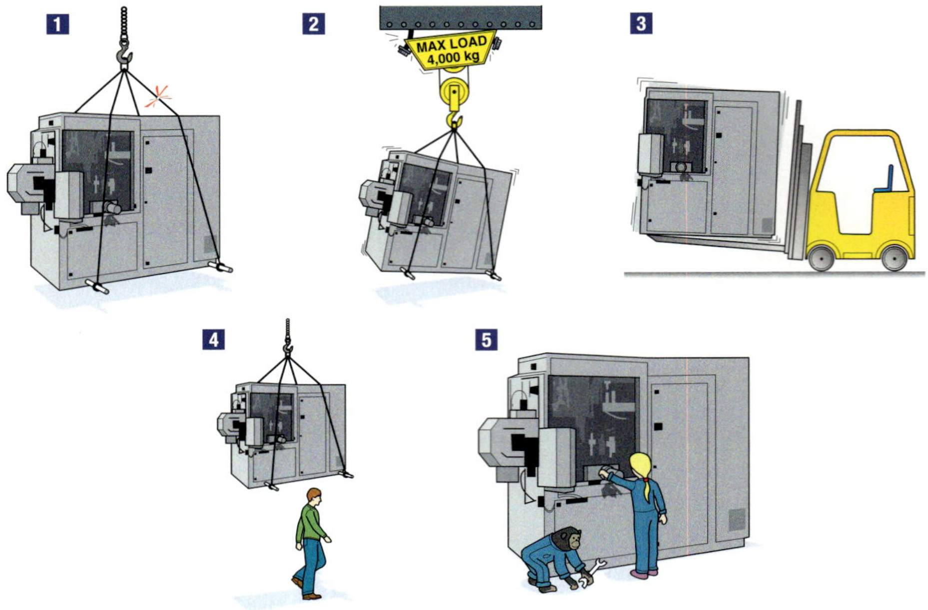


## 13 Preparations for installing the machine

13

**Before the tube production system can be installed, PP-Services in Lahr, Germany has to prepare the facilities for the compressed air supply. Alain phones Lahr. Listen, and answer these questions.**

1 What is holding up the installation?
2 When will the technicians arrive?
3 If the machine is not installed to schedule, what will happen?
4 How does Alain react to the delay?

## 14 Apologizing and giving explanations

**Which of the phrases (1–5) do you use for functions a–e, below?**

1 We'll send out a replacement immediately.
2 I can understand that you're upset.
3 We've had some problems with our …
4 I really must apologize about this.
5 Please don't worry.

a offering an apology
b giving an explanation
c calming down the customer
d showing the customer that you know how he/she feels
e reassuring the customer that the mistake will be rectified

## 15 A rude phone call

**a Alain phones a German company, RR Transportmaschinen, to complain about missing pages in a manual. Janine Rolff, who answers the phone, is not very polite. Read the conversation and decide with a partner how you would improve it. (Tip: some of the phrases below may help you.)**

MS ROLFF Rolff.
ALAIN Oh. Hello! Erm, do you speak English?
MS ROLFF Yes.
ALAIN This is Alain Dufour calling from Orbis in Colmar, France. Is that RR Transportmaschinen?
MS ROLFF Yes.
ALAIN And, erm, who am I speaking to, please?
MS ROLFF Rolff.
ALAIN Ah. Ms Rolff. Is Mr Kahn in the office, please?
MS ROLFF No. He's not. He's in a meeting.
ALAIN Could you give him a message, please?
MS ROLFF Yes.
ALAIN Could you tell him that there are some pages missing in the manual which he sent along with the RR45.
MS ROLFF OK. He'll send you a new manual.
ALAIN Thanks. But is there any chance you could fax or email me the missing pages first? We can't use the machine, you see.
MS ROLFF Which pages?
ALAIN It's pages 18 to 21. Listen, I ...
MS ROLFF What's your fax number?
ALAIN Well, my fax number is ...

- *Good morning, Mr Dufour.*
- *If you give me your fax number, ...*
- *I'll arrange for a new manual to be sent out to you straight away.*
- *Janine Rolff. Sales department.*
- *No, I'm afraid he's not. He's in a meeting. Can I take a message?*
- *Oh, dear. I'm sorry to hear that.*
- *Oh, of course. Which pages are missing exactly?*
- *Pages 18 to 21. Right. I'll fax these pages to you straight away.*
- *RR Transportmaschinen, Rolff. Guten Morgen.*
- *What time would you like Mr Kahn to call you?*
- *Yes, I do. How can I help you?*

14 **b Now listen to the phone call. Did Ms Rolff's improvements match your suggestions?**

## 16 Over to you

**Work with a partner. You are going to practise apologizing and giving explanations. When you have finished, change roles and repeat the exercise.**

Partner A: Look at File 4 on page 90. Partner B: Look at File 10 on page 94.

Grammar

## *If* sentences types 1 and 2

**1** If you **can** get it done in the next two hours, we **will** still **be able to** set up the TET.

**2** We **will have** enormous production problems if you **don't manage** to get the work done.

**3** If we **had to**, we **would work** through the night.

**4** If it **came** to that, I **would be** very angry.

**5** We **could run** into trouble if there **was** a delay in getting the machine installed.

- *If*-Sätze vom Typ 1 drücken eine erfüllbare Bedingung aus. Beachten Sie die Zeitenfolge: Der *if*-Nebensatz steht im *simple present*, im Hauptsatz steht das *will future*. (1, 2)
- *If*-Sätze vom Typ 2 drücken eine möglicherweise erfüllbare, meist unwahrscheinliche Bedingung aus. Beachten Sie die Zeitenfolge: Der *if*-Nebensatz steht im *simple past*, im Hauptsatz wird *would* + Infinitiv verwendet (3, 4), wobei *would* durch *could, might* oder *should* ersetzt werden kann. (5)
- Steht der *if*-Nebensatz an erster Stelle, wird er vom Hauptsatz durch ein Komma getrennt. (1, 3, 4)

## 17 Practice 1: *If* sentences type 1

**Complete the sentences using the correct form of the verbs in brackets.**

1. If unauthorized personnel … (carry out) the installation, they … (not observe) the safety regulations.
2. If the equipment … (fall), it … (endanger) the safety of anyone standing underneath.
3. If the technician … (not connect) the wiring correctly, someone … (get a shock).
4. The technician … (have) a nasty accident if he … (not tie back) his hair.
5. If the technicians … (lift) the machine in a clumsy way, they … (damage) the components.
6. If the machine … (arrive) early tomorrow, the foundations … (not be) ready.
7. … (it/upset) the project if we … (reschedule) installation for next week?
8. … (forklift / tip over) if the driver … (go) too fast?

## 18 Practice 2: *If* sentences type 2

**Describe the following chain reaction using *if* sentences.**

Example:
*If Jo used the wrong cable, the insulation would melt.*
*If the insulation melted, …*

- Jo / use / wrong cable
- insulation / melt
- circuit breaker / trip
- current / switch off
- machines / not work
- production / stop
- boss / be angry

*If the forklift was driven too fast, …*

# Extra material

## Difficulties with an installation

**Last month KLP Textiles, Finland ordered a machine from Jahn GmbH, Freiburg. Unfortunately, the machine could not be installed to schedule. The CEO of KLP has asked for a report. Read the report and do the tasks below.**

### Difficulties with the installation of the MMM3

**The following is the background to our recent difficulties with the installation of the MMM3 ordered by KLP Textiles, Finland. We could not install the machine on time for a number of reasons, only some of which resulted from our actions.**

### Failures on our part

Due to problems with one of our vehicles, the machine was dispatched five hours late from our factory. Under normal circumstances this would not be a problem. We had planned in extra time to cover possible delay in delivery and would have been able to make up the time and hit the deadline. Due to extreme weather conditions, however, the machine was delivered over nine hours behind the scheduled time.

While we have to take responsibility for the problems with the vehicle, this was not the main reason that the deadline for installation of the MM3 was missed. KLP have to share some of the responsibility for this.

### Failures on KLP's part

As soon as we became aware that bad weather was forecast, we asked KLP if we could postpone delivery of the machine by two days. KLP insisted that we deliver and assured us that if there was a delay, they would offer us their assistance to keep the work on schedule by careful planning on their part. They failed to do this in two ways.

First, delivery had to be made via a very narrow road at the rear of KLP's factory. We arranged with KLP that this road would be kept clear to allow access for our vehicle. Despite the fact that KLP were informed of the possible delay, the road was kept clear for only thirty minutes after the scheduled delivery time; no plans for clearing the road in the event of a delay had been made. This led to our driver having to wait a further ninety minutes for the road to be cleared again.

Second, KLP failed to keep the access routes inside their factory clear. As a result of this, we had to wait for KLP employees to move equipment before we could take the MMM3 to its location in the factory. This resulted in a further delay of two hours.

## 1 Working with words

**Find words in the text with the same meaning as these German words.**

**1** rechtzeitig
**2** verspätete Lieferung
**3** den Abgabetermin einhalten
**4** Verantwortung übernehmen für
**5** verschieben
**6** geplant, planmäßig
**7** Zugangswege
**8** Standort

## 2 Mediation

**Work with a partner. Take turns to summarize what the text says in German.**

## 3 Whose fault is it?

**Work in groups and discuss who is responsible for the failure to install and commission the MMM3 on time. Give reasons for your opinions.**

# 6 Parts of a machine

## 1 Warm-up

**a** **Where do you see pneumatic and hydraulic systems in use in everyday life? Look at these photos of parts of machinery, equipment, etc. and decide with a partner what the things are. Then say which of them work with pneumatic and which with hydraulic systems.**

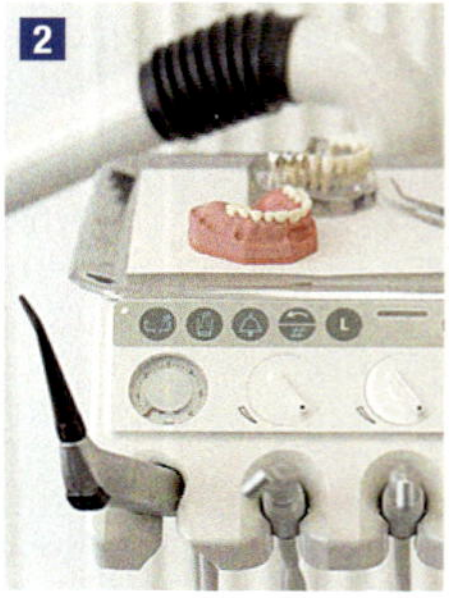

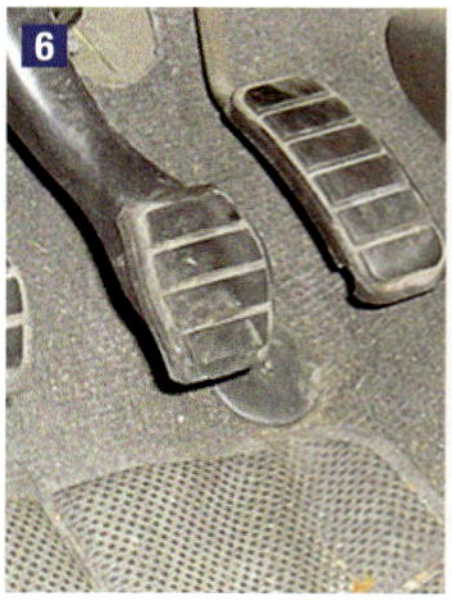

**b** **Take turns to explain how pneumatic and hydraulic systems work.**

## 2 Arranging a maintenance check

15

A few months ago, PackSys Global (Switzerland) Ltd installed a tube production system for Orbis, Colmar (see Unit 5). Isabell Frey, one of the mechatronics technicians who was involved in the installation of the machine, gets this note from her boss and phones Alain Dufour, the production manager at Orbis, to make an appointment for the first maintenance check.

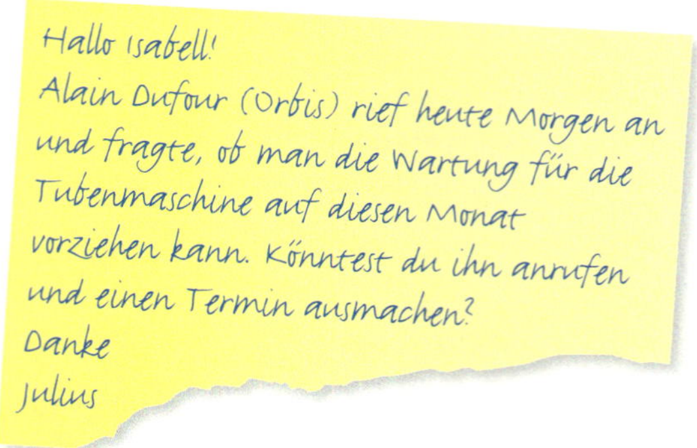


**Read or listen to the telephone conversation then do the exercise on the opposite page.**

**ALAIN** Bonjour. Orbis, Colmar. Puis-je vous aider?

**ISABELL** Erm, hello. Is that Mr Dufour? Do you speak English?

**ALAIN** Yes, Alain Dufour speaking. How can I help you?

**ISABELL** It's Isabell Frey from PackSys Global here.

**ALAIN** Ah, Ms Frey. Thanks for calling.

**ISABELL** My pleasure, Mr Dufour. I understand you'd like to have your scheduled maintenance check on the tube production machine carried out a month earlier than the date stated in the warranty.

**ALAIN** That's right. There's a big order coming up next month and it would be great if the machine could be thoroughly checked beforehand. As a matter of fact, the machine's not working at full efficiency so it would be good if a maintenance inspection could be carried out as soon as possible.

**ISABELL** Oh. It's not working properly? What's the problem?

**ALAIN** Well, the machine functions well for a bit, then it starts to slow down. The tubes aren't loading at full speed. We're getting error message number 17. I've had a look in the handbook, but don't really understand what it's all about.

**ISABELL** Hm. It sounds to me as if there's a problem with an actuator or a sensor. Whatever it is, it'll definitely have to be looked at. When would you like us to come out?

**ALAIN** Well, we can't afford any downtime this week. Would it be possible for you to come out at the beginning of next week?

**ISABELL** Next week. ... OK. Let me just have a look at the calendar to see what our work schedule looks like ... Monday afternoon is clear. Would Monday at 2.30 suit you?

**ALAIN** Sorry, I can't make it then. I'm afraid another appointment has been scheduled for that time. Can I suggest Tuesday morning instead – say at 10 o'clock?

**ISABELL** Yes, that would be fine. I'll probably be accompanied by Julius Keller.

**ALAIN** Splendid. I look forward to seeing you both again. Till Tuesday, the 4th November at 10.

**ISABELL** I look forward to seeing you again, too, Mr Dufour. Bye for now.

**ALAIN** Goodbye.

## 3 Working with the text

**a Julius has asked Isabell for feedback on the telephone call. Write a short report in German.**

**b Now answer these questions in your own words.**

1 Why would Mr Dufour like to have the maintenance check carried out a month earlier than planned?
2 What is the problem with the machine?
3 How urgent is the problem?
4 Which components might be causing the problem?
5 When do Mr Dufour and Isabell agree to meet?
6 Who will be with Isabell when she visits Mr Dufour's company?

## 4 Working with words

**a Find the English for the following in the dialogue on page 57.**

1 Wartung
2 Garantie
3 Fehlermeldung
4 Aktor
5 Ausfallzeit
6 Terminplan

**b Match six of the definitions (a–h) to the words above. Two definitions do not match.**

a a note on a display which tells you that something is wrong
b a device which can react to light, heat, etc.
c the time during which a machine or a computer is not working
d a plan that lists all the things you have to do and when they have to be done
e doing well with no waste of time or money
f keeping something in good condition by checking or repairing it regularly
g an element which is driven by compressed air in a pneumatic system
h a written promise given by a company that something will be replaced or repaired if there is a problem within a particular period of time

**c Now find the words in the text which match the two extra definitions.**

Learning

### Wie man Fachvokabeln mit Hilfe des Internets übersetzt

Wenn Sie eine Fachvokabel im zweisprachigen Wörterbuch nicht finden, können Sie natürlich in einem Online-Wörterbuch wie www.leo.org danach suchen.

Werden Sie dort nicht fündig, sollten Sie nicht gleich aufgeben, sondern es mit ein paar Kniffen versuchen, z.B.:

- Einen deutschsprachigen Wikipedia-Artikel aufrufen und auf die englische Version umschalten.
- Bei Google das Wort durch Eingabe von Kontextwörtern (d.h. Wörtern, die normalerweise zusammen mit dem gesuchten Wort auftreten) suchen. Sie können auch versuchen, das englische Wort zusammen mit deutschen Kontextwörtern zu suchen.
- Technische Seiten wie www.howstuffworks.com aufrufen, den Artikel für das Oberthema suchen und die englisch bezeichneten Teile anhand der Abbildungen erkennen.

Sie möchten z.B. wissen, was „Entlötlitze" auf Englisch heißt. Rufen Sie den Artikel über Löten in Wikipedia auf und wechseln Sie zur englischsprachigen Seite. Dort werden in dem Abschnitt über *desoldering* (Entlöten) auch *desoldering wicks* erwähnt. Die Google-Suche nach „desoldering wicks entlötlitze" bestätigt, dass es sich um das gesuchte Wort handelt.

## 5 Asking about a problem

**You get this note from your boss. Role-play the phone call with a partner.**

Carlos García Fernández von Torres Motor Systems SL hat angerufen. Er möchte, dass wir uns den Extruder an ihrer Verpackungsmaschine PKS-90T mal ansehen. Kannst du ihn bitte anrufen und fragen, worum es geht? Ggf. einen Termin ausmachen.

## 6 Confirming the appointment

**Later in the day, Isabell writes an email to Alain Dufour confirming their appointment. Complete the email with seven of these words.**

accompanied • confirm • dear • forward • greetings • honoured • mentioned • pleasure • wishes

New | Reply | Reply All | Forward | Flag | Print | Delete | Send & Receive | Address Book | Find

...[1] Mr Dufour

It was a ...[2] to speak to you on the telephone today. I am writing to ...[3] our appointment on 4 November at 10.00 a.m.
As I ...[4] on the telephone, I will be ...[5] by my supervisor, Julius Keller.
I look ...[6] to seeing you on 4 November.

Best ...[7]
Isabell Frey

## 7 Making and confirming appointments

**Find suitable endings (a–k) for the sentence beginnings. (There are two endings too many.)**

**1** Could you possibly come → **i**
**2** I'm afraid I can't make
**3** Hold on. I'll just check
**4** My colleague will be
**5** Would 12.15
**6** Sorry, but I have
**7** I look forward to
**8** Could I suggest
**9** I'll see

**a** suit you?
**b** my diary.
**c** it at that time.
**d** about next week?
**e** an alternative date?
**f** coming along, too.
**g** meeting you on Wednesday.
**h** check our work schedule.
**i** on Friday afternoon instead?
**j** you on Monday, 12th October at 3 p.m.
**k** another appointment on that day.

## 8 Over to you

**Work with a partner and make arrangements to meet.**

Partner A: Look at File 6 on page 91. Partner B: Look at File 11 on page 95.

# 9 Electric motors

The electric motors in the tube production machine need to be looked at during the maintenance check.

**a** **Discuss in groups what you know about electric motors.**

**b** **Read the text below, then summarize the information in German.**

http://www.knowitall.com/electricmotors

Apple .Mac Amazon eBay Yahoo! News ▾

Knowitall.com

## Electric motors

The basic scientific principles behind the electric motor are that like poles repel and unlike poles attract, and that electric current flowing through a conductor produces a magnetic field that has both a north and a south pole. An electric motor is essentially a rotating electromagnet. As the electromagnet rotates, electrical energy is transformed into mechanical energy, which can then perform work.

There are many different designs for electric motors, but these are the three basic types:

### 1 DC motor

In a DC motor, the electromagnet (the armature) rotates within a permanent magnetic field created by two opposing magnets (the exciter field). The interaction of the north and south pole of the armature and the north and south poles of the exciter field produces a torque, which rotates the armature. A device which is called a split ring commutator reverses the magnetic poles of the armature as it rotates. Carbon brushes 'brush' against each half of the split-ring commutator and conduct the electrical energy to the armature. The reversal of the armature poles as they interact with the poles of the exciter field alternately pushes and pulls the armature around.

### 2 AC motor

A typical AC motor consists of an outside stationary stator having coils supplied with AC current to produce a rotating magnetic field, and an inside rotor attached to the output shaft. The rotor and shaft are given a torque by the rotating field.

In an induction AC motor, the rotor rotates more slowly than the magnetic field produced by the electrical supply. Often the rotor is made up of two rings, one at either end, connected by solid bars. Washing machines, dishwashers, etc. use induction motors.

In a synchronous AC motor, the rotor rotates at the same speed as the rotating magnetic field. This type of motor can also be used as an alternator. Very large synchronous motors are used in the French high-speed train, the TGV.

### 3 Universal motor

Universal motors may use AC or DC supply current. Unlike DC motors, universal motors have to use an electromagnet to create the stator field, because the field has to reverse its polarity every half cycle of the AC current.

The advantage of the universal motor is that it has the useful characteristics of a DC motor, e.g. high starting torque and very compact design, but can be used with an AC power supply. Unlike AC motors, universal motors can easily exceed one revolution per cycle of the mains current. This makes them useful for appliances such as blenders, vacuum cleaners, and hair dryers where high-speed operation is desired.

*Sources: www.bergen.org and http://en.wikipedia.org*

# 10 Working with words

**Find the English words in the text above.**

1 (sich) abstoßen
2 (sich) anziehen
3 (sich) drehen
4 Drehmoment
5 Stromwender
6 Bürsten
7 Spulen
8 Läufer
9 Hauptwelle
10 Polarität
11 übersteigen
12 Netzstrom

## 11 Which type of motor is it?

**Which two types of motors described on the web page are shown here?**

## 12 Electric motor rating plates

**Work with a partner. You are going to look closely at the rating plate on one of the electric motors in the tube production machine.**

**a Match the English words (1–7) with their German translations (a–g).**

1 rating plate
2 rated speed
3 power factor
4 star connection
5 delta connection
6 mains frequency
7 mechanical performance

a mechanische Leistung
b Dreieckschaltung
c Sternschaltung
d Leistungsfaktor
e Nenndrehzahl
f Netzfrequenz
g Typenschild

**b Now say what a–g in the following rating plate are.**

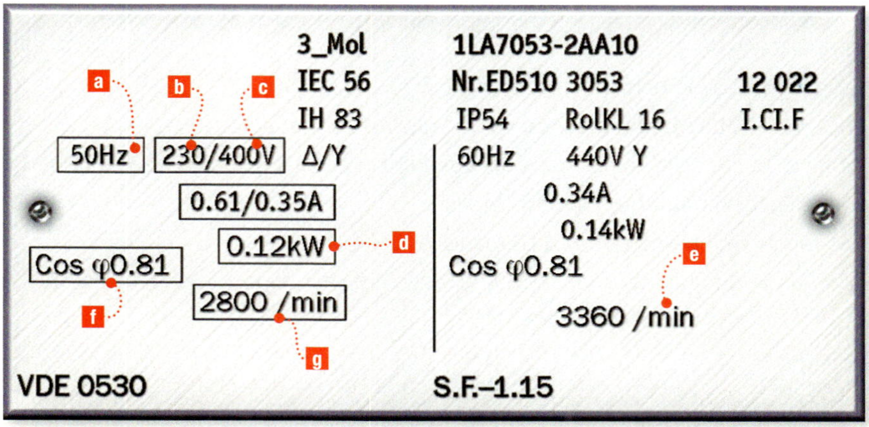


1 voltage for delta connection
2 the motor's power factor
3 the motor's mechanical performance
4 voltage for star connection
5 the rpm (revolutions per minute)
6 the motor's rated speed
7 the mains frequency required to run the motor

## 13 Overview of a pneumatic system

**Isabell looks over some of her training notes about pneumatics. Match the words in *italics* with the correct parts of the diagram (a–h).**

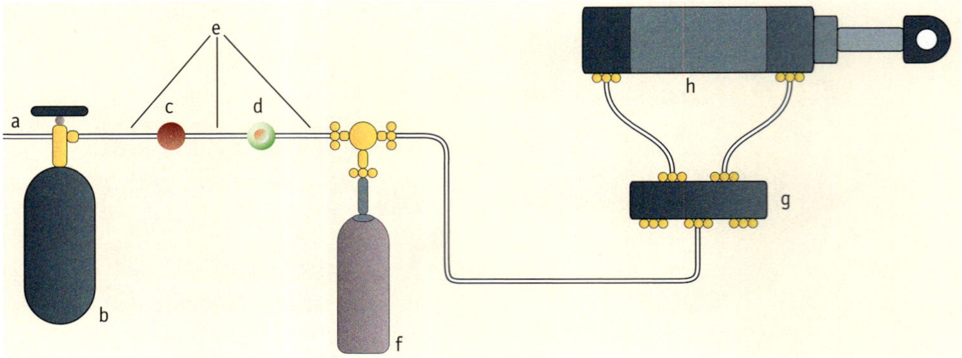


1 The *control valve* controls the release of air or gas to the pneumatic actuator.
2 After it has left the storage tank but before it reaches the regulator, compressed air is cleaned by being passed through a *filter*.
3 In some pneumatic systems, compressed air or gas is held in a *buffer tank* before it reaches the control valve. This helps to ensure that constant air pressure is kept up.
4 The *regulator* controls the pressure of the air or gas which is fed to the actuator allowing the system to run efficiently at all times.
5 Compressed air or gas is transported through the system by means of *pneumatic hoses*.
6 The compressed air comes into the system through the *inlet for compressed air*.
7 The *actuator* is powered by all of the components in the service unit. It is the business end of the pneumatic system which performs the required industrial task.
8 Air or gas is held in the *storage tank*.

Grammar

### The passive

1 Compressed air **is held** in a buffer tank.
2 The machine **has been checked** twice since it **was installed**.
3 When all of the safety measures **had been checked**, the machine **was switched on**.
4 Every machine **must be handled** carefully.
5 The repairs **will be carried out by** Isabell and Julius.

- Das Passiv wird häufig verwendet, wenn man auf eher unpersönliche Art und Weise über Fakten, Vorgänge und Zahlen berichtet.
- Das Passiv wird mit der entsprechenden Form von *be* + Partizip Perfekt (3. Form des Verbs) gebildet. Wie im Aktiv können alle Zeitformen gebildet werden.
- Das Passiv mit Modalverben wird mit dem Modalverb + *be* + Partizip Perfekt gebildet. (4)
- Da im Passiv die beschriebene Handlung wichtiger ist als die Person/Sache, die sie ausführt, kann der/die Handelnde weggelassen werden. (1–4)
- Soll der/die Handelnde genannt werden, benutzt man das Wort *by* (= von, durch). (5)

## 14 Practice

**Here is a page from the operating manual of the tube production machine. Use the information to write passive sentences.**

Example: *The electric cabinet should be cleaned out with a vacuum cleaner once a week.*

### 6.2 Maintenance and lubrication diagrams

#### 6.2.1 Cleaning and checking jobs

| Nr. | Position | Maintenance function | Interval |
|---|---|---|---|
| | **Electric cabinet** | | |
| 11 | - Cabinet | Clean out with a vacuum cleaner. | Weekly |
| 12 | - Ventilator | Change or blow out the filter mats with compressed air. | Monthly |
| 13 | - Electronic cards | Blow out carefully with low-pressure compressed air. | Every half year |
| | **Water** | | |
| 14 | - Pipes, hoses, nipples, valves | Check for water leakage. | Weekly |
| 15 | - Filters | Check.<br>Replace the filter cartridge if required. | Every two weeks |
| | **Pneumatics** | | |
| 16 | - Pipes and fittings | Check for any air leakage. | Weekly |
| 17 | - Filters | Clean the filter. | Weekly |
| | **Hydraulics** | | |
| 18 | - Pipes, fittings and valves | Check for oil leakage.<br>If any leakage is found, tighten the fittings. | Weekly |

## 15 How a pneumatic system operates

**a Match the verbs (1–6) with the words on the right (a–f) to make phrases. (Sometimes there is more than one possibility.)**

| | |
|---|---|
| **1** release | **a** air to the necessary pressure |
| **2** pass | **b** compressed air through the system by means of hoses |
| **3** hold | **c** air to the pneumatic actuator |
| **4** keep up | **d** the pressure |
| **5** transport | **e** compressed air through a filter |
| **6** compress | **f** air in a tank before it reaches the control valve |

**b Now use the phrases to make complete passive sentences about how a pneumatic system works.**

Example: *In a pneumatic system, air is released to ...*

## 16 A hydraulic system

**a The tube production machine also features a hydraulic system. Match the words in the box with the parts labelled in the photograph.**

accumulator • electric motor • heat exchanger • oil level indicator • sensor/actuator box • tank • thermostatic valve • valve block

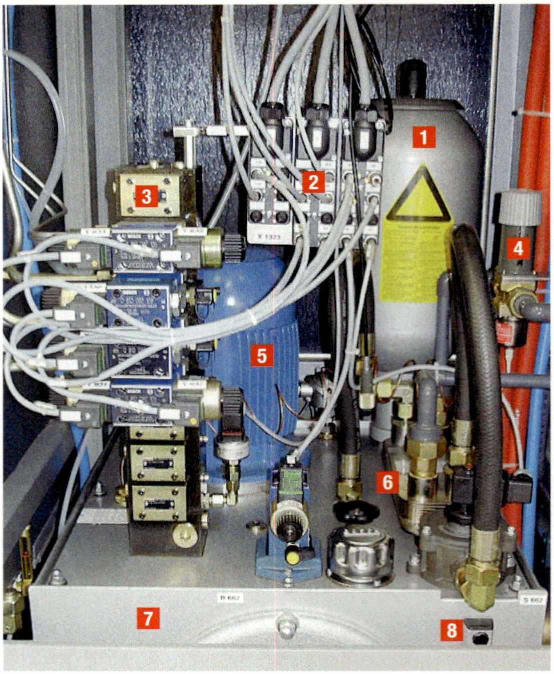


**b Listen as Isabell explains the various parts of the hydraulic system to Alain and check your answers.**

**c Isabell talks about the sensor/actuator box. What does it do?**

## 17 The accumulator

**a Say which components of the accumulator (1–5) the words in the box refer to.**

bladder • connection with the hydraulic system • mushroom valve • nitrogen • oil

**b Match these two sentences with two of the diagrams (a–c).**

1 The pressure in the system is stabilized, the oil flows out of the accumulator via the mushroom valve and the nitrogen expands to fill the accumulator bladder.
2 The pressure within the system is normal and the nitrogen-filled bladder fills the accumulator.

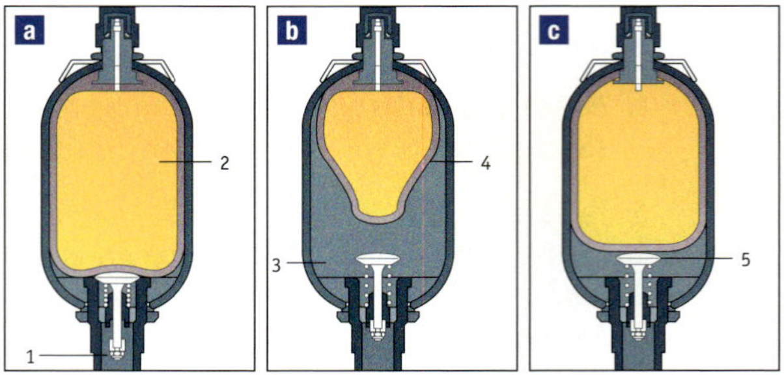


**c Now write a description of the remaining diagram. Use the following phrases.**

- shock in the system / pressure / build up
- mushroom valve / open
- allow / oil / flow into / accumulator
- nitrogen / compressed / smaller volume

# Extra material

## Maintenance, Repair and Overhaul

ROBOTS AND OTHER TYPES of factory plant are used to churn out products with short lifespans. When the CD player which you bought three years ago breaks down, you probably won't send it in for repair. Instead, you're likely to scrap it and buy a new one. The plant used to make these disposable products is a different matter, however. It's hugely expensive and companies have to look after it to make sure that it continues to work and that they get a return on investment.

This is what lies behind the strategy of MRO (Maintenance, Repair and Overhaul). Rather than scrapping industrial plant such as production robots after a few years, increasing numbers of manufacturers are using MRO strategies to squeeze as much profitable use as possible from their capital investments.

Rolf Moll, a consultant who helps companies to realize the full potential of their plant by using MRO, says that the idea has become so popular that major robot manufacturers have turned over large parts of their businesses to servicing customers' MRO requirements.

'The idea of MRO is self-explanatory,' says Moll. 'You keep costs down by making the most of what you already have. There are basically two approaches in an MRO programme,' he goes on. 'Equipment can either be remanufactured or recertified. Remanufacture – this means that the equipment is maintained and repaired by a technician working on site; recertified – that means that the manufacturer of the equipment takes it back and gives it a complete overhaul.'

The potential cost savings of MRO are obvious. But there are other less obvious savings that can be made.

'Using MRO as a starting point, a company can buy overhauled plant in order to duplicate an existing production line that it set up in the past. This duplication cuts out the need for investment in staff retraining,' Moll explains.

Of course mechatronics technicians play a key role in any MRO strategy. And as MRO becomes increasingly popular, so the role of mechatronics frontliners will become even more important.

### 1 Find expressions in the text which match these definitions.

1 large machinery in a factory
2 to dispose of a piece of machinery or equipment which no longer works
3 a product which is made to be thrown away after use
4 companies which build or make things
5 pieces of equipment or buildings which require a large amount of money to be invested in them
6 in the factory, workshop or wherever the equipment is located
7 a product passes along this, parts are made, put together, etc.
8 very important

### 2 List the advantages of MRO given in the text. Do you agree with the writer? Why / Why not? Add other advantages you can think of.

### 3 Which types of maintenance are carried out in your company? Describe a typical job.

7

# A presentation about a PLC

## 1 Warm-up

**Work with a partner.**

**a** **Match the cartoons (1–4) with the descriptions of things that can go wrong during a presentation (a–d).**

**a** Addressing a formal audience too informally.
**b** Using too many slides.
**c** Not being prepared; being disorganized.
**d** Nervous body language; making hand movements which distract the audience.

**b** **Talk about other things which people get wrong when giving a presentation. What would you do differently?**

## 2 The introduction to a presentation

17

**Gordon Trip from Sonic Services Ltd is giving a presentation to a group of German mechatronics trainees. Read or listen to the introduction to his presentation.**

'Good morning, everybody. My name's Gordon Trip. I'm the PLC Training Co-ordinator here at Sonic Services. This morning, I'm going to tell you about the S7-200

Programmable Controller. This is one of the most successful compact programmable logic controllers or PLCs for short. My apologies for giving this presentation in English, by the way, but your English is much better than my German and Mr Bryan, your English teacher, tells me that this will be good practice for you.

First, I'd like to tell you about the structure of my presentation. I'm going to begin by giving you an overview of the S7-200: its key components, capabilities and applications. I think the S7-200 is one of the most flexible PLCs on the market today and I'll be telling you why.

The second section of my presentation will focus on the installation of the S7-200. I'll outline the installation procedures and the system requirements. I'll also look at installation guidelines for the S7-200 expansion module.

Then, thirdly, I'll turn my attention to the installation and operation of the S7-200-Micro/WIN Software. We'll look here at the software's key features and applications. After that, I'll take a look at basic programming for the S7-200. In the next section of my presentation we'll look at network communications and how to connect the S7-200 to ensure compatibility with different types of networks.

Finally, I'll deal with what you can do if you have a problem with the S7-200 by taking a look at troubleshooting procedures.

Now, as you know, we have a hands-on session this afternoon where we'll look closely at installation procedures and guidelines for the S7-200. This will involve working with the installation instructions and the PLC unit itself. The aim of my presentation this morning is to prepare you for the afternoon session. You'll really start to understand the system when you get your hands on it.

Right, let's get started. I plan to cover a lot in the next fifty minutes and I'm sure you'll find it all very interesting. I'll be happy to answer any questions at the end of my presentation, of course.

I'll start now by looking at my first topic. ...'

## 3 Working with the text

**Put the subjects below in the order Gordon Trip plans to talk about them.**

**a** Basic programming for the S7-200.
**b** Installing the software; description of key features and applications.
**c** Network compatibility and installation.
**d** Troubleshooting and problem-solving.
**e** Overview of the key components, capabilities and applications of the S7-200.
**f** Installation of the S7-200 and guidelines for installing the expansion module.

## 4 Working with words

**Work with a partner.**

**a Match these words and phrases from the text with their German translations (a–h).**

1 capabilities
2 applications
3 system requirements
4 guidelines
5 compatibility
6 troubleshooting
7 installation instructions
8 hands-on

a Installationsanweisungen
b Richtlinien
c Fehlerbeseitigung
d praktisch, aktiv
e Systemvoraussetzungen
f Fähigkeiten, Leistung
g Kompatibilität
h Anwendungen

**b Complete the sentences with six of the English words above.**

a Steffie really enjoys … . She's good at finding and fixing problems.
b Please read the … . You must know and understand the rules before you start.
c These computers can't be used together. There's no … between them.
d I prefer to do something rather than talk about it. I'm a real … person.
e Set up the equipment using the … .
f What are the … of this machine?

**c Which two words have not been used in exercise b? Write two sentences using these words. Read them out to the class.**

## 5 Introducing a presentation

**a Gordon Trip's colleague is also giving a presentation. Match the beginnings (1–9) with the endings (a–i) to make sentences which she uses to introduce her presentation.**

1 Good afternoon. I'm Joan Barrow. I'm involved → e
2 Today, I'll be telling you
3 To start with, I'm going to give you an outline
4 Then, we'll have a look
5 After that, I'll go
6 Moving on, I'm going to talk you
7 To end with, I'll be dealing
8 Later today, you'll be able to work
9 Should there be any questions

a of my presentation.
b with problems that may come up with the S7-200.
c about the S7-200 PLC.
d at how the S7-200 is installed.
e in PLC training here at Sonic Services.
f over the S7-200's capabilities.
g about my presentation, I'll deal with them towards the end.
h with the PLC yourselves.
i through the software, programming and so on.

**b Find similar sentences in the text on pages 66–67.**

## 6 Over to you

**Work with a partner.**

**a** **Imagine that you are going to give a presentation about a piece of equipment you use. Use phrases from exercises 5a and 5b to prepare the introduction.**

**b** **Now take turns to give the introduction to each other.**

Learning

### Tipps zur Körpersprache während einer Präsentation

- Verschränken Sie nicht die Arme – das kann abweisend wirken.
- Vermeiden Sie nervöse Handbewegungen. Es wirkt unruhig, wenn Sie sich an der Nase kratzen oder mit den Fingern durchs Haar fahren.
- Stellen Sie mit den Zuhörern Augenkontakt her, aber konzentrieren Sie sich dabei nicht auf eine einzelne Person. Gute Redner lassen ihre Blicke durch den Raum schweifen und scheinen dabei jeden im Publikum anzuschauen.

Üben Sie Ihre Körpersprache vor einem Spiegel. Dann fühlen Sie sich bei einer Präsentation sicher, weil Sie wissen, dass Sie sich überzeugend mitteilen können.

## 7 Presentation slides

**Match four of the titles (a–f) to the Powerpoint slides (1–4). Two titles do not match.**

**a** Installation configuration
**b** Troubleshooting
**c** Presentation overview
**d** Practice programming for the S7-200 Micro PLC
**e** Major components of the S7-200 Micro PLC
**f** Customer service and the S7-200

**1**
- An S7-200 Micro PLC consists of a CPU module alone or with a variety of optional expansion modules.
- The CPU module combines a central processing unit (CPU), power supply, and discrete I/O points in a compact stand-alone device.

**2**
- You can mount a CPU either on a panel or on a standard rail.
- You can mount the CPU horizontally or vertically.
- An I/O expansion cable is also available to add flexibility to your mounting configuration.

**3**
- Overview of the S7-200
- Looking at installation procedures
- Getting started with programming

**4**
- Making the best use of programming FAQs
- Defining the automated task before programming
- Monitoring the results of the practice program

## 8 Presentation bullet points

**Work with a partner.**

**a Use these words to complete the presentation bullet points below.**

accommodate • bus • cable • clearance • cooling • devices • executes • expansion • fault • field • modules • port • power • status • task • vertically

- The CPU …[1] the program and stores the data for controlling the automation …[2] or process.
- The CPU and expansion modules are designed for natural convection …[3]. You must provide a …[4] of at least 25mm, both above and below the units, for proper cooling.
- The …[5] supply provides electrical power for the base unit and for any …[6] module that is connected.
- If you are installing a CPU horizontally or …[7] on a panel, you must allow 75mm for the minimum panel depth.
- The inputs and outputs are the system control points: the inputs monitor the signals from the …[8] devices (such as sensors and switches), and the outputs control pumps, motors or other …[9] in the process.
- If you plan to install additional …[10] horizontally or vertically, allow a clearance of at least 25mm on either side of the unit for installing and removing the module. The extra space is required to engage and disengage the …[11] expansion connector.
- The communications …[12] allows you to connect the CPU to a programming device or to other devices.
- …[13] lights provide visual information about the CPU mode (RUN or STOP), the current state of the local I/O, and whether a system …[14] has been detected.
- Be sure to allow enough space in your mounting design to …[15] the I/O wiring and communication …[16] connections.

**b Sort the bullet points into topic A, Installation configuration and topic B, Major components of the S7-200 Micro PLC and take turns to explain them in German.**

## 9 Checking information

**a Listen to the question and answer session at the end of Gordon Trip's presentation. Who asks about what?**

connecting PCs • inputs and outputs • installation • powerpoint slides • the hands-on session

1 Jenny Sager    2 Berndt Anders    3 Stefan Braun

**b Listen again and choose the best ending for each sentence. Only one ending is possible.**

1 Inputs are
  **a** connected to outputs.
  **b** the pieces of information sent by the sensors to the CPU.
  **c** sensors which measure the outputs and send the data to the CPU.

2 The information sent to the CPU is used to control the
  **a** machine.
  **b** outputs.
  **c** sensors.

3 When you use a PC with the S7-200, you
  **a** only need a communication processor card.
  **b** need both a PC/PP1 cable and a communication processor card.
  **c** need either a communication processor card with a multipoint interface cable or a PC/PPI cable.

4 When you install additional modules, you should leave 25mm clearance around the
  **a** CPU.
  **b** bus expansion connector.
  **c** extra modules.

5 You should do this so that there is space
  **a** to work with the modules.
  **b** for the CPU.
  **c** to engage and disengage the bus expansion connector.

**c Listen again. Which of these phrases do you hear?**

1 Would you like to ask me something?
2 I was a bit confused about ...
3 Are you saying ...?
4 May I ask something?
5 Anyone else?
6 Could you clarify what you said about ...?
7 Do you mean ...?
8 Does that answer your question?
9 I'd just like to clear up a point about ...
10 If I understood you correctly, ...
11 Will there be time for questions later?
12 There will be more time for questions later.

**d Put the phrases into two lists: presenter and member of the audience.**

## 10 Checking and clarifying

**a Rearrange the phrases to form questions which can be used to check spoken information and to ask for clarification.**

1 with pneumatic and hydraulic systems / that / you don't like working / do you mean
2 next week / correctly / we should start learning / if I understand you / about PLC programming
3 saying that / the manual properly / he didn't read / are you
4 what / we should spend / by saying that / doing hands-on stuff / do you mean / more time
5 do more work / are you saying that / on it / he should

**b Here are four statements which need to be checked and clarified. Match them with four questions from 1–5 above.**

a He should have read the installation instructions before he started.
b We have to spend more time on hands-on activities.
c We'll look at installation this week. We'll cover the rest later.
d I enjoy working with PLCs and sensors but I'm not sure about the other parts of the production line technology.

**c Which question does *not* answer a statement? With a partner, think up an appropriate statement for it.**

Grammar

### The *-ing* form and infinitive

1 This will **involve working** with the PLC unit itself.
2 I **plan to cover** a lot in the next fifty minutes.
3 I'll **start looking** at my first topic now.
4 You'll really **start to understand** the system.
5 Can we **begin by looking** at inputs and outputs?
6 It's **a waste of time trying** to set up the equipment without reading the instructions.

- Die *-ing*-Form folgt auf Verben wie *enjoy*, *imagine*, *involve*, *stop*, *suggest*. (1)
- Auf andere Verben wie *agree*, *choose*, *decide*, *expect*, *hope*, *learn*, *plan*, *prefer*, *want* folgt der Infinitiv. (2)
- Auf einige Verben wie *begin*, *continue*, *hate*, *like*, *love* und *start* kann sowohl die *-ing*-Form als auch der Infinitiv folgen. (3, 4)
- Wenn auf eine Präposition ein Verb folgt, muss man die *-ing*-Form verwenden. (5)
- Nach bestimmten Ausdrücken steht auch die *-ing*-Form (6). Beispiele:

*He's **good at giving** presentations.*

*I'm **afraid of making** a mistake.*

***How about going** to the cinema after school?*

## 11 Practice

**Complete the sentences.**

1 Could you imagine … (work) abroad?
2 She enjoys … (learn) about computer programming.
3 Our company has decided … (buy) one of the latest models.
4 The next step in the installation procedure involves … (check) the wiring.
5 Could you go on with … (prepare) the presentation, please?
6 It's a waste of time … (ask) questions before the end of the presentation.

## 12 PLC operating cycle

**a Complete the text and the flow chart below with the words in the box.**

activated • consist • determine • execution • input • instruction • operates • recorded • scan • stages • understanding

A PLC …[1] by constantly *scanning* a program. This …[2] cycle can be envisaged as consisting of three critical …[3]. In reality, there are normally more than three but let's look at the important stages to gain a general …[4] of how PLCs work. Normally the other stages …[5] of monitoring the system and taking care of current internal counter and timer value updates.

*Stage 1*

**CHECK INPUT STATUS**

The PLC examines each input to …[6] if it is switched on or off. Is the sensor connected to the first input on? What is the status of the second and third inputs? This data is …[7] in the PLC's memory ready for use during the next stage.

*Stage 2*

**EXECUTE PROGRAM**

The PLC runs the programs one …[8] after the other. Perhaps the program states that if the first input was on, then the first output should be activated. Since the system can already see which inputs are on from the previous stage, it will be able to decide whether the first output should be …[9] based on the state of the first input. The results of this program …[10] are stored for use in the next stage.

*Stage 3*

**UPDATE OUTPUT STATUS**

The PLC system updates the status of the outputs. The outputs are updated based on which inputs were activated during the first stage and the results of the program run during Stage 2. Based on the example in Stage 2, the PLC would now activate the first output on the basis that the first …[11] was on: the program stated that the first output should be activated when this condition is fulfilled.

**b Summarize the PLC operating cycle in German.**

# 13 Installation instructions

**Which of these subjects are covered in the manual on the opposite page?**

1 Wiring requirements for the S7-200
2 Description of how to mount the S7-200
3 Execution times for STL instructions
4 Procedure for connecting the expansion module
5 Health and safety advice
6 Description of input/output control
7 Troubleshooting steps for failed installation of the expansion module
8 Description of how to position the expansion module in relation to the CPU module

# 14 Important verbs

**a Find verbs in the text with the same meaning as the German verbs below. Copy the English verbs into your exercise book.**

1 sich vergewissern
2 abschalten
3 befestigen
4 öffnen
5 einhängen, festhaken
6 entfernen
7 hineinstecken
8 aufhebeln
9 vorgehen
10 beschädigen
11 ausrichten
12 schieben

**b Can you remember the phrases in the text that the verbs are used with? Write them beside the list of verbs in your exercise book, then check in the text.**

# 15 What is happening?

**a Work with a partner.**
**Which installation procedure in the manual does this picture illustrate?**

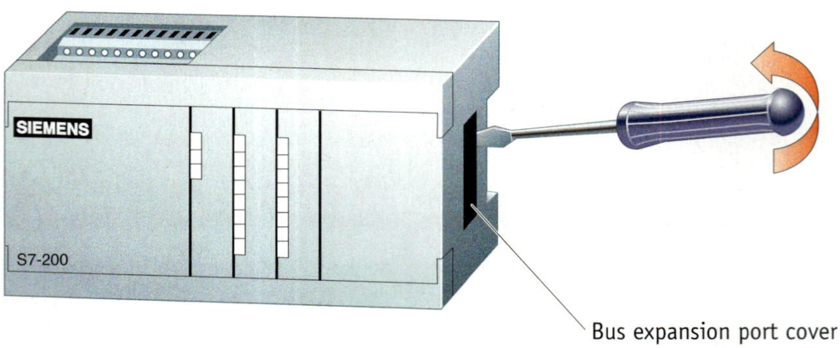


**b Now cover the manual and describe the procedure in the picture in your own words.**

**c Summarize the information in the manual in German.**

# Installing an S7-200 Micro PLC onto a Standard Rail

***Warning***

*Before installation always ensure that power to the S7-200 modules is disabled. Death or serious personal injury and/or damage to equipment may result from failure to disable all power to the S7-200 modules and related equipment during installation.*

Follow these steps to install the CPU module:

1. Secure the rail every 75mm to the mounting panel.
2. Snap open the clip (located on the bottom of the module) and hook the back of the module onto the rail.
3. Snap the clip closed, carefully checking to ensure that the clip has fastened the module securely onto the rail.

**Note**

Modules in an environment with high vibration potential or modules that have been installed in a vertical position may require standard rail stops.

If you are installing an expansion module, follow these steps:

1. Remove the bus expansion port cover from the existing module housing. (Insert a screwdriver into the space between the bus expansion port cover and the housing and pry the cover away gently.) Ensure that the plastic connecting joints are completely removed. Proceed with caution so as not to damage the module.
2. Insert the bus connector into the bus expansion port of the existing module and ensure that the connector snaps into place.
3. Ensure that the expansion module is correctly oriented with respect to the CPU module. If you are using an expansion cable, orient the cable up towards the front of the module.
4. Snap open the clip and hook the back of the expansion module onto the rail. Slide the expansion module onto the bus connector until it snaps into place.
5. Snap the clip closed to secure the expansion module to the rail. Check carefully to ensure that the clip has fastened the module securely onto the rail.

## 16 Presentation equipment

**Look at the pictures (1–8) and say what the equipment (a–h) is called.**

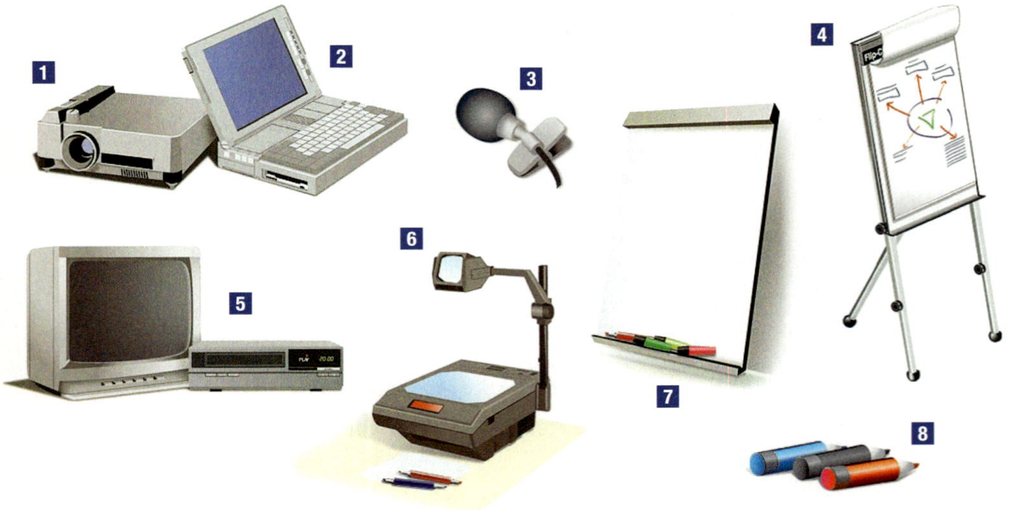


**a** overhead projector (OHP)
**b** TV and video
**c** laptop
**d** clip-on microphone
**e** whiteboard
**f** projector
**g** flipchart
**h** marker pens

## 17 Clear signposting phrases for presentations

**Sort the phrases below into their respective categories.**

**a** Main section **b** Summary **c** Conclusion

1 At this point I would like to say something about …
2 To go over the main points once again …
3 To end with, I'm going to take a detailed look at …
4 I'd just like to review what I've said …
5 I'd like to close by saying …
6 Moving on to my next section, …

## 18 Over to you

**Work in groups.**

**a** **Write a presentation about the installation procedures described in the manual on page 75.**

**b** **Give your presentation to the class. The other students will listen and make notes. Answer their questions at the end of the presentation.**

Information

A presentation needs a clear introduction and a summary at the end. Use signposting phrases (see exercise 5 on page 68 and exercise 17 above).

# Extra material

## Ladder logic for PLCs

**A** Ladder logic or the ladder programming language is a method of drawing electrical logic schematics. It is a graphical language which was originally invented to describe logic made from relays. The name is based on the observation that programs in this language resemble ladders, with two vertical 'rails' and a series of horizontal 'rungs' between them.

**B** A program in ladder logic, also called a ladder diagram, is similar to a schematic for a set of relay circuits. Because of this similarity, ladder logic is useful because a wide variety of engineers and technicians can understand and use it without much additional training.

**C** Ladder logic is widely used to program PLCs, where sequential control of a process or manufacturing operation is required. It is useful for simple but critical control systems, or for reworking old hardwired relay circuits. As PLCs become more sophisticated, ladder logic is used in more complex automation systems.

Most vendors of PLCs also provide associated ladder logic programming systems.

**D** Ladder logic is best thought of as a set of closely related programming languages rather than one language. Typically, the ladder logic languages from two manufacturers will not be completely compatible. Even different models of programmable controllers within the same family may have different ladder notation so that programs cannot be seamlessly interchanged between models.

## 1 Keywords

**a Match words in the text to the definitions.**

1 Electrical switches that open and close under control of another electrical circuit.
2 The sides and the 'steps' of a ladder.
3 A drawing of a technical system.
4 Almost the same as.
5 Following in order of time.
6 Complicated and intricate.
7 Work absolutely well and with ease together.
8 Joining up neatly, with no spaces or 'breaks' between one thing and another.

**b Now use your dictionary to translate the words into German.**

## 2 Mediation

**Work with a partner.**
Partner A: Explain paragraphs A and C to your partner in German.
Partner B: Explain paragraphs B and D to your partner in German.

## 3 Alternative PLC programming languages

**Work in groups.**
**Discuss your experience and knowledge of PLC programming languages.**

# 8 Applying for a job

## 1 Warm-up

**Work with a partner.**

**a Match the words with their definitions.**

benefits • location • pay • promotional prospects • responsibilities

**a** the chance to get a better job in the company in the future
**b** how much you money you will earn
**c** the tasks you will perform in the job
**d** where the job is based
**e** extras you may be offered on top of your pay, e.g. sports club membership

**b Which of the factors above are most important to you when you apply for a job? Add some of your own ideas then rank them all in order of importance.**

*The most important thing for me is …*

*I don't care about … but I would like a job which …*

## 2 A job advertisement on the internet

Hannah Berg, a newly qualified mechatronics technician, looks at a website where jobs for technicians are offered. She is interested in the advertisement on the opposite page.

**Read the advert and complete the sentence beginnings.**

1 The company will pay …
2 EOZ AS is looking for somebody who …
3 Candidates need to be …
4 The company's policy is to …

Learning

### Effiziente Suche nach spezifischen Angaben beim Lesen

Es gibt verschiedene Arten, einen Text zu lesen. Texte können nach bestimmten Informationen durchsucht oder ganz genau gelesen werden. Welche Technik Sie wann anwenden, hängt jeweils von Ihrer Absicht ab. Bei der Suche nach spezifischen Angaben sollte man sich nicht von der Fülle der übrigen Informationen ablenken lassen. Wenn es Ihnen beispielsweise bei der Durchsicht von Stellenausschreibungen darauf ankommt, sich über die benötigten beruflichen Qualifikationen zu informieren, können Sie selektiv lesen und sich auf Angaben wie *ability*, *experience*, *knowledge* und *qualification* konzentrieren.

http://www.jobs-for-technicians.com/sensortechnician

Apple Mac Amazon eBay Yahoo! News

Home More Jobs

# Sensor Technician

**Pay in the region of €32,000 (according to experience and age) + benefits**

EOZ AS is a fast-growing manufacturing company specializing in the production of medical scanning equipment based in the Republic of Latvia. We are seeking a motivated Sensor Technician with a background in mechatronics to support the team at our high-tech production facility in Riga.

As the successful candidate, you will have the ability to work on your own initiative and to plan and carry out testing and monitoring of our production line sensors. You should be a good team player who can think on his/her feet and troubleshoot effectively in the event of an emergency. Additionally, you must be able to work under pressure while still having an eye for detail. Finally, you need to be a good communicator in order to liaise effectively with our manufacturing equipment suppliers.

**Requirements:**

- Accredited qualification in mechatronics
- 2–3 years of work experience in a production environment
- Ability to demonstrate practical understanding of sensor applications
- Experience in monitoring and troubleshooting for sensor systems
- A full, current and clean driving licence
- Good knowledge of English

**Closing date for applications: 3rd November 20..**

To apply for the job send your CV together with a covering letter and the names of two referees to:
*Agnese Daknys, HR Assistant, EOZ AS, Riga, Jurkalnes 57, LV 1134.*

Or email your application and CV to agnese_daknys@EOZ-HR.lv

*EOZ AS is an equal opportunities employer. We select candidates for jobs according to their skills and experience, and regardless of their gender, ethnicity, sexual orientation, religion or age.*

## 3 Working with the text

**Read the advertisement again and make two lists of what the company is looking for.**

| Skills and experience | Personal qualities |
|---|---|
| *– ability to plan and carry out testing* | *– motivated* |

## 4 Working with words

**Work with a partner.**

**a Match the German words and expressions (1–12) with English words and expressions from the text.**

1 ansässig in
2 eine Ausbildung in etwas haben
3 unterstützen
4 Fähigkeit
5 unter Druck arbeiten
6 mit jemandem zusammenarbeiten
7 Berufserfahrung
8 Verständnis
9 Bewerbung
10 Lebenslauf
11 Anschreiben
12 Geschlecht

**b Choose four of the words or expressions and describe them in English to your partner. Can he/she guess which words you are describing?**

## 5 Talking about your interests and abilities

**Look back at the small website on page 78 and answer these questions.**

1 Which of the jobs would you be interested in?
2 Which (extra) qualifications might you need for the job?

## 6 Finding equivalents

**It is sometimes difficult to find the English equivalent for German types of schools and qualifications. Match the German words on the left with the English equivalents on the right.**

1 Grundschule
2 Haupt-/Realschule
3 Ausbildung
4 Gymnasium
5 Mittlerer Schulabschluss
6 Abitur
7 Fachhochschulreife
8 Fachhochschule
9 Berufschule / Berufskolleg / Berufsbildende Schule
10 Duales System
11 Zeugnis
12 Gesamtschule

**a** vocational college/school
**b** primary (GB) / elementary (US) school
**c** German equivalent of GCSE (General Certificate of Secondary Education) (GB)
**d** German equivalent of A level (GB) / High School Diploma (US)
**e** entrance qualification for universities of applied science
**f** apprenticeship scheme
**g** comprehensive school (GB)
**h** secondary (GB) / high (US) school
**i** vocational training
**j** report
**k** grammar (GB) / high (US) school
**l** university of applied science

## 7 Getting jobs abroad

**a Read the text on the opposite page. Use your dictionary to check any words you do not know, then summarize the text in German.**

http://www.jobexpert-uk.ac.uk/

Apple .Mac Amazon eBay Yahoo! News

## Applying for jobs in different countries

In most countries, you should not include copies of your school or vocational college certificates or references with your application. You may be asked to produce these at a later date, for example, after you have been offered the job.

Include the names and contact details of two referees at the end of your curriculum vitae (CV), one of whom should be your current employer.

Unless asked to, don't sign and date your CV and don't include a photograph of yourself with your application. This is particularly the case in the USA where employment laws against racial discrimination state that photographs are not to be used.

Under 'Personal details' provide your full name, contact details and date of birth.

Provide details about your education, training and work experience in reverse order: under each of these headings, start with the present and work your way back through your personal history.

Don't forget to include a section called 'Other skills and experience' in which you can list things which don't fit in other parts of the CV. This might be your language skills or whether or not you can drive.

Include details of hobbies and pastimes on your CV.

Potential employers often provide an application form for job candidates instead of asking for a CV. If this is the case, don't enclose a CV, but try to include all the details you want to mention on the form.

**b Now make a list of the things which are different from Germany. Compare your list with your classmates' lists.**

## 8 Hannah's CV

**Work with a partner. Hannah decides to apply for the job at EOZ AS. Copy the headings (1–7) from her CV, then use the tips on the website to organize her points (a–l) under them.**

1 Personal details
2 Education and qualifications
3 Training
4 Work experience
5 Other skills and experience
6 Hobbies and pastimes
7 Referees

a Mr Achim Kuhl, Production Training Manager, Schnelling AG, Bonn, Germany
b 20.. Abitur (equivalent to A Levels) Max Roth Gymnasium, Bonn, Germany
c Reading, aerobics
d Summer 19.. Youth Summer Camp helper in Madison, Wisconsin, USA
e European Union driving licence
f Marital Status – single
g 20.. – 20.. Apprenticeship as Mechatronics Technician – Schnelling AG, Bonn, Germany
h Inline skating
i Date of birth 17.04.19..
j Nationality German
k Summer, 20.. Work experience placement at Schnelling AG, Bonn, Germany
l March – June 20.. Volkshochschule (Adult Education College) Bonn: course in rhetoric and communication

## 9 Over to you

**Write your CV in English.**

## 10 Hannah's covering letter

**a Complete Hannah's letter with the sentence endings (a–k). One of the endings does not fit.**

Dear Ms Daknys

**Application for the post of sensor technician with EOZ AS**

I am writing to express …[1]. I would very much like to …[2]. Please find my curriculum vitae enclosed. As you will see, I am …[3]. I have just …[4].
I am applying for this job because I would …[5]. In addition, I think that the post with EOZ AS would be an …[6].
I feel that I am well suited to the post as I am a hardworking person and …[7].
Moreover, I am sure that my extensive training …[8].
I have tried to provide all the information …[9]. If you need further details, …[10].
I look forward to hearing from you soon.

Yours sincerely

Hannah Berg

Hannah Berg

- **a** highly motivated
- **b** be considered for the job
- **c** completed my apprenticeship
- **d** a qualified mechatronics technician
- **e** please do not hesitate to get in touch
- **f** interesting challenge for me at the beginning of my career
- **g** and experience will ensure that I am effective in the role
- **h** like to gain experience of working in another country
- **i** and keen to advance in my career in mechatronics
- **j** that you might need in my curriculum vitae
- **k** my interest in the post of Sensor Technician with EOZ AS

**b Now make a sentence using the remaining ending. Where could it fit in Hannah's letter?**

## 11 Personal qualities

**a How would you express the ideas in the box in German?**

fast learner • good team player • hardworking • keen to learn new skills • responsible • thorough

**b Now use them (and other phrases you know) to describe your personal qualities.**

## 12 Over to you

**Write a covering letter applying for this job.**

**a First, make notes describing**

- your skills and experience,
- your personal qualities,
- why you would like to get the job.

**Mechatronics technician** wanted for an international production project. Based at our main production facility in Wiesbaden.

**GLOBALTECH AG**

- Recently qualified
- Flexible
- Willing to travel
- Good knowledge of English
- Hard worker
- Willing to work shifts

*CV and covering letter to:*
Ms Georgina Freedman
HR Co-ordinator
Globaltech AG

**b Now write the covering letter. Use your notes to say why you are the right person for the job.**

## 13 Hannah's interview at EOZ AS

19

**a Work with a partner. Which questions would you expect to be asked at a job interview? Make a list.**

*When did you …*

*How long have you …*

**b Now listen to the interview. How many of your questions do you hear? (You needn't have written down the exact words.)**

**c Listen again, take notes and answer the questions below.**

1. Why did Hannah decide to become a mechatronics technician?
2. What did Hannah do after she left school?
3. What did she learn about sensors during the first year of her apprenticeship?
4. At what stage in her apprenticeship did she learn about troubleshooting for sensor systems?
5. Did Hannah learn about monitoring and maintenance for sensors
   a. in a specific year of her training?
   b. throughout her apprenticeship?
6. What are Hannah's strengths and weaknesses?

Culture

### Strengths and weaknesses

In job interviews in English-speaking countries it is common to be asked what you think your strengths and weaknesses are. Not surprisingly, interviewees often present 'weaknesses' which are really strengths, e.g. 'I'm afraid I'm a bit of a perfectionist'. Even though perfectionism is considered by some people to be a negative trait, employers expect you to do a good job, so this is probably one of the qualities they are looking for.

When you prepare yourself for an interview, don't forget to think about your strengths and weaknesses.

Grammar

## The simple past and present perfect

**1** I **wanted** to do something practical when I left school.
**2** Why **did** you **decide** to become a mechatronics technician?
**3** We'**ve learned** a lot about sensors in the last three years.
**4** I'**ve looked** for a job on the internet but I **haven't found** anything yet.

### Simple past

Diese Zeitform verwendet man, um abgeschlossene Handlungen und Ereignisse in der Vergangenheit auszudrücken. (1, 2)

### Present perfect

Diese Zeitform bezieht sich auf Ereignisse oder Handlungen, die in der Vergangenheit begonnen haben und bis zur Gegenwart andauern oder deren Auswirkungen noch in der Gegenwart sichtbar oder spürbar sind. (3, 4)

Bestimmte Zeitangaben verlangen gezielt und ausschließlich nach dem *simple past* bzw. dem *present perfect*:

*Simple past* – Zeitadverbien wie *2 hours ago*, *when I left school*, *last week* oder *in 2005*, die sich nur auf die Vergangenheit beziehen.

*Present perfect* – Zeitadverbien wie *since 2005*, *in the last 3 years*, *already*, *recently* oder *for the last 2 hours*, da der Zeitraum, auf den sich der Satz bezieht, bis zur Gegenwart reicht.

## 14 Practice

**Complete the sentences using either the simple past or the present perfect form of the verbs in brackets.**

**1** I ... (be interested in) mechatronics since I left school.
**2** I ... (see) the advertisement on your website last week.
**3** We ... (be) very impressed with what you ... (say) at the interview yesterday.
**4** I ... (start) my apprenticeship as a mechatronics technician two years ago.
**5** How long ... (you / work) for Bognor Ltd now?
**6** She ... (not / work) since she finished her apprenticeship in April.
**7** He ... (finish) his training three weeks ago.
**8** In the past we ... (have) six new apprentices every year, but since 2004 we ... (take on) ten new apprentices per year.

## 15 Interview questions and answers

**Complete these sentences with suitable words and phrases.**

**1** Where ... you ... about our company?
**2** Why ... you ... to become a mechatronics technician?
**3** What ... you ... about sensors during your training?
**4** In the first year of my training we ...
**5** During the last three years I ...
**6** How long ... the company been in business?

## 16 A job interview

**Work with a partner and role-play a job interview.**

Partner A: Look at File 5 on page 90. Partner B: Look at File 12 on page 95.

## 17 The test

**Hannah got top marks for the test she took after her interview. Do the test yourself. How many points can you get?**

**A Match these German words with words in the text below. (10 points)**

1 aktiver Raum
2 Tastkopf
3 erfassen
4 Schaltabstand
5 Lichtstrahlen
6 abgeben
7 unterbrochen
8 Endlagenschalter
9 Schließerkontakt
10 Wechselkontakt

**a** This type of sensor gives off an electronic signal when metal components are in the active zone of the sensor head. Switching unit clearance is up to 30 mm.

**b** This type of sensor detects the presence of all materials such as plastic, wood, glass and liquids. Switching unit clearance is up to 30 mm.

**c** This type of sensor emits light beams which are either reflected back to the receiver or sent direct to the receiver. In both cases, the sensor detects the presence of an element when the signal is interrupted or broken.

**d** When this type of sensor is used as a limit switch with pneumatic cylinders, it will normally have open contacts or change-over contacts. The sensor is activated when a piston fitted with a permanent magnet passes under it.

**B Which four of these types of sensors (1–5) do the texts above refer to? (4 points)**

1 Inductive proximity sensors
2 Capacitive proximity sensors
3 Optoelectrical sensors
4 Temperature sensors
5 Magnetic sensors

**C Describe the fifth type of sensor in a similar way. (4 points)**

**D Choose a, b or c to complete these sentences. (2 points)**

1 The most common cause of failure in a light sensor is that

- **a** there is dirt or dust on the sensor.
- **b** the sensor is located in an area where there is too much light.
- **c** the sensor is located in a machine which is operating at too high a temperature.

2 A magnetic sensor usually fails because the sensor

- **a** is operating in a very humid area.
- **b** is too far away from the object it is meant to detect.
- **c** is too close to the object it should detect.

# 18 Teamwork test

**After she has done Parts A–D of the test, Hannah has to do a practical task in a group session. Work in groups of three and do Part E on the test sheet below.**

## E The technical data on this specification sheet have got mixed up. Work together to find out what 1–25 refer to.

**SIEMENS**

**SIGUARD Light Barrier 3RG7822**

| Technical data |
|---|
| Type designation |
| Operating range |
| Range adjustment |
| Light type |
| Reflection angle |
| Minimum object size |
| Housing dimensions |
| Housing material |
| LED status display<br>3RG7822-7C000 (receiver)<br>3RG7822-7B600 (transmitter) |
| Connection type |
| Max. cable length |
| Operating voltage $U_B$ |
| No-load running |
| Switch output Q and (pollution) |
| Output current |
| Voltage drop at switch output |
| Switch frequency (ti/tp 1:1) |
| Test input |
| Response time |
| Electr. protection class |
| Safety circuits |
| Operating ambient temp. |
| Storage/Transport temp. |
| Protective system |
| Weight |

| | | |
|---|---|---|
| 1 | ABS, impact proof | |
| 2 | prot. against incorrect connection, prot. against short circuit | |
| 3 | using sensitivity setter (2 turns with display) | |
| 4 | ≤ 200 mA | |
| 5 | < ± 2° | |
| 6 | infra red, 880 nm, pulsed | |
| 7 | –20 … +80°C | |
| 8 | 100 m | |
| 9 | ≤ 30 mA | |
| 10 | PNP (n.o.) | |
| 11 | ≥ 13 mm Ø | |
| 12 | 3RG7822-7BG00(transmitter) / 3RG7822-7CD00 (receiver) | |
| 13 | 700 Hz | |
| 14 | 2 | |
| 15 | 0 … +60°C | |
| 16 | IP 67 | |
| 17 | 50 mm × 50 mm × 17 mm | |
| 18 | approx. 60 g/pair | |
| 19 | 0.7…15 m (sensitivity adjustable) | |
| 20 | t on = t off ≤ 0.7 ms | |
| 21 | 10…30 V DC, max. residual ripple 10% | |
| 22 | ≤ 2.4 V | |
| 23 | unit plug M 12×1, 4-pole, rotating | |
| 24 | red on: | light beam broken |
| | green on: | light beam free |
| | flashing mode | contaminated, misaligned, |
| | red green: | distance between transmitter and receiver too great, sensitivity set too low |
| | red on: | transmitter off |
| | green on: | transmitter on |
| | orange on: | transmitter on (operation via 3RG7827-1DE2) |
| 25 | < 2 V or open: | transmitter = off |
| | > 10 V: | transmitter = on |

Language

## Words and phrases used in teamwork

How should we go about this?

Maybe we should try a process of elimination.

The weight will be expressed in grams.

What do you think the … might be?

OK, now let's look at the …

# Extra material

## 1 Finding a job abroad

**Read the text. Put paragraphs a–e in the correct logical order then explain each paragraph in a single sentence in German.**

http://www.jobs_advice.com/

Apple .Mac Amazon eBay Yahoo! News

### Top tips for working abroad

Working abroad for two or three years can be good for your career. You'll get an insight into another culture and gain experience which could be invaluable later in life. However, *getting* that job abroad can be quite difficult. Here's a step by step guide about how to do it.

**a** The next step is to find that job! Look on the internet at jobsites, check out the jobs pages of newspapers in the country in question.

**b** Once you have secured your job offer, work permit and visa it's time to make sure that you have the right health insurance and pension cover to work in the country. The immigration services of the country in question should be able to help you with this.

**c** But before deciding that you would like to work in a country, it's important that you visit it first. You can only find out if a particular country is right for you if you spend some time there finding out what the people are like. It's especially important to find out what the cultural differences between your own country and the country you want to live in are.

**d** If you are thinking about working in another European Union country, you have the right to enter and live in the country in order to find work. However, countries outside of the EU will generally require you to apply for a visa before entering the country as a resident.

**e** Once you've decided that you would like to spend some time working in another country, you should do some research into whether you can realistically expect to secure a job there. The embassy or consulate of the country in question may also be able to provide you with information about your chances of finding employment.

## 2 Understanding the text

**Read the text again and complete these sentences in your own words.**

1 Working in another country can help you …
2 Before you start to look for a job abroad, you must …
3 When you've got a job in another country, you …

## 3 Working in Germany

**Work with a partner.**
**What information and advice about working in Germany would you give to someone from another country? Make a list. Compare your information and advice with the rest of the class.**

– working hours
– attitudes towards small talk
– smoking in the workplace

APPENDIX

# Pairwork files

## File 1 Unit 2, Exercise 9

### Partner A

1 You are a new apprentice at Plastico plc. You are in the HR building and you want to know how to get to the engineers' shop and the sports club. Ask for directions.

Follow your partner's directions on the plan and say which number is the engineers' shop and which is the sports club.

2 You are an engineer at Plastico plc. You are in the warehouse. Give the new apprentice directions when he/she asks for them.

(The training centre is number 2 on the plan and the canteen is number 3.)

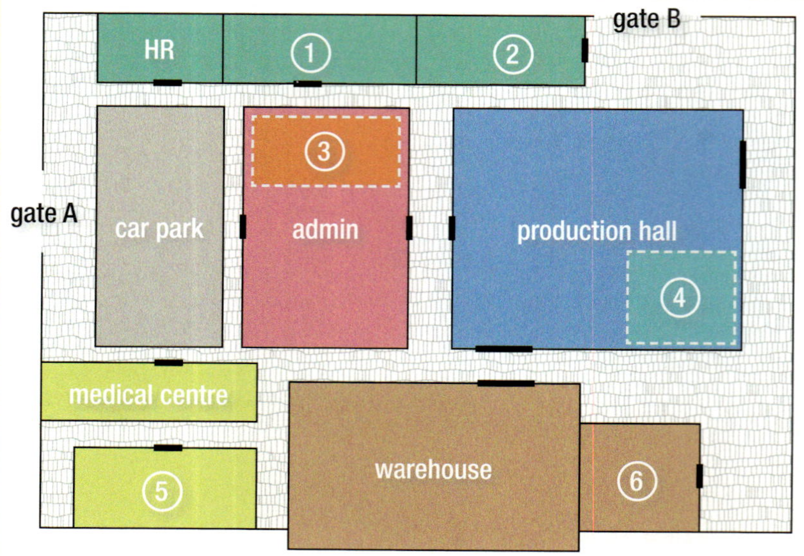


## File 2 Unit 3, Exercise 9

### Partner A

Read out these mathematical equations. Partner B will write them down. When you have finished, check that Partner B has written them correctly.

1 $A > B$

2 $e = mc^2$

3 $\sqrt{9} = 3$

4 $30 + 13 = 43$

5 $2 \times 2 \times 2 \times 2 = 2^4$

6 $324 - 78 = 246$

7 $9{,}842 \times 55 = 541{,}310$

8 $0.65 \div 5 = 0.13$

## File 3 Unit 3, Exercise 4

### Partner A

You have the information you need to answer your partner's questions.

What information do you need from him/her to complete your specifications sheet?

*Examples:*

**1** What is the approximate input impedance on the Z-axis input?

**2** Can you tell me the sensitivity of the vertical system, please?

FEATURES: 20 MHz, Dual Channel High sensitivity 1mV / DIV TV Synchronization Z axis input ALT Triggering Function CH1 output

| | | |
|---|---|---|
| CRT | Type | 6-inch rectangular type with internal graticule 8 x 10 DIV (1 DIV = 1 cm) |
| | Z-axis Input | Input Impedance: approx. ...[1] |
| | | Sensitivity: above 5Vp-p |
| | | Bandwidth: DC ~ 2MHz |
| VERTICAL SYSTEM | Sensitivity | ...[2] |
| | Bandwidth | DC (AC 10Hz) ~ 20MHz (-3dB) |
| | | DC (AC 10Hz) ~ 7MHz (-3dB) x 5MAG |
| | Rise Time | ...[3] |
| | Input Impedance | approx. 1MΩ |
| | Input Coupling | AC, CD, GND |
| | Vertical Mode | ...[4] |
| HORIZONTAL SYSTEM | Sweep Time | 0.2μS ~ 0.5S/DIV±3%<br>100nS ~ 50mS/DIV±5% (x 10MAG)<br>20nS ~ 50nS/DIV uncalibrated |
| TRIGGER | Trigger Mode | AUTO, NORM, TV-V, TV-H |
| | Trigger Source | CH1, CH2, ALT, LINE, EXT |
| | Trigger Coupling | AC |
| | Trigger Slope | "+" or "-" |
| X - Y OPERATION | Sensitivity | ...[5] |
| | X-axis Bandwidth | DC ~ 500kHz |
| | Phase Error | 3° or less from DC ~ 50kHz |
| OUTPUT SIGNAL | Trigger Signal Output | Voltage: approx. ...[6] |
| | Calibrator Output | 1kHz Squarewave, ...[7] |
| POWER SOURCE | | AC115V/230V ± 10%, 50/60Hz |
| ACCESSORIES | | Power cord x 1 Instruction manual x 1; Probes (10:1/1:1) x 2 |
| DIMENSIONS & WEIGHT | | ...[8] |

## File 4 Unit 5, Exercise 16

### Partner A

1 You are the production manager at Glass Engineering, Glasgow. A technician from WG GmbH in Germany should have come to your factory yesterday to oversee the installation of a machine. The machine has arrived, but there is no technician. Phone Mr/Ms Drews at WG and ask for an explanation. Make notes before you start. Include the following points in your phone call:

- your name (you can use your own name or make one up);
- say that the machine arrived yesterday, as arranged;
- ask what has gone wrong;
- say that you have a tight schedule; the machine must be installed and working by a certain date (e.g. tomorrow / the end of the week);
- ask Mr/Ms Drews what he/she is going to do.

2 You are in charge of Sales and Service at Krüger Robotics GmbH. Last week, one of your mechatronics technicians was sent to R & R Automobiles, Coventry to install and commission one of your pick and place robots. When he/she got back to Germany, the mechatronics technician reported that the robot had a mechanical fault. The production manager at R & R Automobiles is going to phone you to ask why no replacement part has arrived. Make notes before you start. Include the following points in your phone call:

- your surname is Stein;
- apologize to the production manager;
- assure him/her that a replacement has been ordered;
- explain why the replacement has not arrived (e.g. problems with your suppliers / transport strike);
- tell the production manager that you will do your best to get the replacement part installed and the robot commissioned as soon as possible.

## File 5 Unit 8, Exercise 16

### Partner A

You are going to interview Partner B.

You would like to know:

- why Partner B decided to train as a mechatronics technician;
- how long he/she has been doing his/her apprenticeship;
- about the technological areas Partner B has covered during his/her training;
- what Partner B's strengths and weaknesses are;
- why he/she would like the job.

Conclude the interview by asking Partner B if he/she has any questions about the job or the company.

When you have finished the interview, swap roles and do it again. Use File 12 on page 95.

## File 6 Unit 6, Exercise 8

### Partner A

| | |
|---|---|
| **MONDAY 4**<br>8:00 – 11:00 Visit to Nanotar GmbH | **FRIDAY 8**<br>9:00 – 10:30 English course<br>13:00 – 17:00 Maintenance checks at Nanotar |
| **TUESDAY 5**<br>13:00 – 17:00 Repair job at Profistar AG | **SATURDAY 9**<br>9:00 – 17:00 Seminar – Customer relations<br>JFK Building, Seminar Room 5 |
| **WEDNESDAY 6**<br>9:00 – 13:00 Visiting Deutsche Textile GmbH | **SUNDAY 10** |
| **THURSDAY 7**<br>9:30 – ?? Dentist<br>13:00 – 17:00 Training session on pneumatics | |

## File 7 Unit 2, Exercise 9

### Partner B

**1** You are an engineer at Plastico plc. You are in the HR building. Give the new apprentice directions when he/she asks for them.

(The engineers' shop is number 4 on the plan and the sports club is number 5.)

**2** You are a new apprentice at Plastico plc. You are in the warehouse and you want to know how to get to the training centre and the canteen. Ask for directions.

Follow your partner's directions on the plan and say which number is the training centre and which is the canteen.

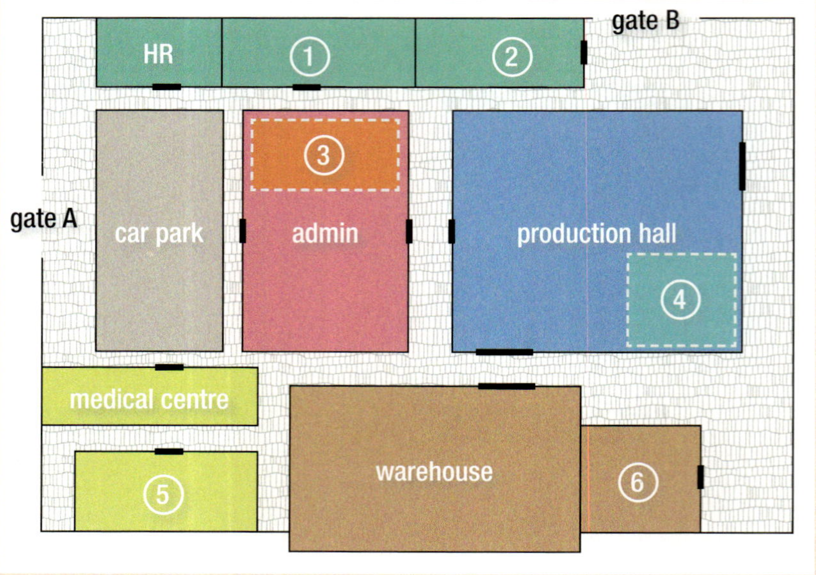


## File 8 Unit 3, Exercise 9

### Partner B

Read out these mathematical equations. Partner A will write them down. When you have finished, check that Partner A has written them correctly.

**1** $7 + 17 = 24$

**2** $3 + 4^2 = 19$

**3** $0.4 \div 0.08 = 5$

**4** $12 \times 12 \times 12 = 12^3$

**5** $4,350 \div 3 = 1,450$

**6** 21 megaohms = 21,000,000 ohms

**7** $\frac{1}{2} - \frac{1}{3} = \frac{1}{6}$

**8** $4,698 \times 67 = 314,766$

## File 9 Unit 3, Exercise 4

### Partner B

You have the information you need to answer your partner's questions.

What information do you need from him/her to complete your specifications sheet?

*Examples:*

**1** Can you tell me about the type of CRT the oscilloscope has, please?

**2** What's the sensitivity of the Z-axis input?

FEATURES: 20 MHz, Dual Channel High sensitivity 1mV / DIV TV Synchronization Z axis input ALT Triggering Function CH1 output

| | | |
|---|---|---|
| CRT | Type | ...[1] |
| | Z-axis Input | Input Impedance: approx. 74kΩ |
| | | Sensitivity: above ...[2] |
| | | Bandwidth: DC ~ 2MHz |
| VERTICAL SYSTEM | Sensitivity | 5mV ~ 5V/DIV ±3%, x 5MAG+5% |
| | Bandwidth | DC (AC 10Hz) ~ 20MHz (-3dB) |
| | | DC (AC 10Hz) ...[3] |
| | Rise Time | 17.5nS (50nS at x 5MAG) |
| | Input Impedance | approx. ...[4] |
| | Input Coupling | AC, CD, GND |
| | Vertical Mode | CH1, CH2, DUAL (ALT/CHOP), ADD, CH2 INV |
| HORIZONTAL SYSTEM | Sweep Time | ...[5] |
| TRIGGER | Trigger Mode | AUTO, NORM, TV-V, TV-H |
| | Trigger Source | ...[6] |
| | Trigger Coupling | AC |
| | Trigger Slope | "+" or "-" |
| X - Y OPERATION | Sensitivity | 5mV ~ 5V / DIV±4% |
| | X-axis Bandwidth | ...[7] |
| | Phase Error | 3° or less from DC ~ 50kHz |
| OUTPUT SIGNAL | Trigger Signal Output | Voltage: approx. 20mV/DIV into 50 |
| | Calibrator Output | 1kHz Squarewave, 2Vpp±2% |
| POWER SOURCE | | ...[8] |
| ACCESSORIES | | Power cord x 1; Instruction manual x 1; Probes (10:1/1:1) x 2 |
| DIMENSIONS & WEIGHT | | 310(W) x 150(H) x 455(D) mm approx. 8kg |

## File 10 Unit 5, Exercise 16

### Partner B

1 You are in charge of the workshop at WG GmbH. The production manager at Glass Engineering, Glasgow is going to phone you to complain. Your firm delivered a machine to Glass Engineering yesterday but the technician who was to oversee the installation of the machine has not arrived. Make notes before the production manager phones. Include the following points:

- your surname is Drews;
- apologize to the production manager;
- explain what has happened to your technician (e.g. flight delayed / is ill);
- tell the production manager that you will do your best to get things back on schedule as soon as possible.

2 You are the production manager at R & R Automobiles, Coventry. Last week, a mechatronics technician from Krüger Robotics GmbH installed one of their pick and place robots in your factory. The robot was installed but could not be commissioned as there was a mechanical fault. The technician said he/she would come back with a replacement part as soon as possible but, so far, no replacement has arrived. Phone Mr/Ms Stein who is in charge of Sales and Service at Krüger's and ask what is happening. Make notes before you start. Include the following points in your phone call:

- your name (you can use your own name or make one up);
- explain what has happened;
- ask what has gone wrong;
- say that you have production dates to meet; the robot must be functioning by a certain date (e.g. tomorrow / the end of the week);
- ask Mr/Ms Stein what he/she is going to do.

## File 11 Unit 6, Exercise 8

### Partner B

| Monday 4 | Friday 8 |
| --- | --- |
| 12:00 – 17:00 Training session on PLC programming | 15:30 Flight to Barcelona |
| **Tuesday 5** | **Saturday 9** |
| 8:00 – 12:00 Visit to Reflector AG | Barcelona |
| **Wednesday 6** | **Sunday 10** |
| 13:00 – 17:00 Repair work at Avelistix GmbH | 19:52 Return flight |
| **Thursday 7** | **NOTES** |
| 9:00 – 11:00 Planning meeting with GA<br>14:00 – 17:00 English training | |

## File 12 Unit 8, Exercise 16

### Partner B

You are going to be interviewed for a job as a mechatronics technician. Think about what you might be asked and make notes.

When you have finished the interview, swap roles and do it again. This time use File 5 on page 90.

APPENDIX

# Transcripts

## 3 Unit 1, Exercise 9

ANNA Right then, Nina, one of the first things you should learn is who's responsible for different sections of the company. Take a look at this diagram and I'll talk you through it.

NINA OK.

ANNA Well, first of all, let's start with Human Resources – HR for short. You've met Marek Sepczuk, he's the HR manager and he's responsible for personnel, pay and training. He's my boss; I report directly to him. Now, on the same level as Marek, we've got Helena Konopka. Ms Konopka's the commercial manager. She's in charge of Accounts. Then there's Ewa Sawicka. She's in charge of Marketing. Ewa's nice. You'll like her.

NINA Good.

ANNA Marketing's basically sales and advertising, but, as we're a small branch, Ewa also deals with customer complaints. Complaints aren't usually part of Marketing.

NINA I imagine it's difficult to be nice when you have to deal with complaints.

ANNA Yes, that's one of the sections where you have to be especially polite to customers. Ewa does a good job, though. She's good at handling complaints from her colleagues, too. Jakub Kolbe, for example, he's the dispatch supervisor. He's in charge of Dispatch. Sometimes things get really hectic down there and Jakub storms upstairs and moans to Ewa about how difficult it all is. She always listens carefully and goes over the problems with him till they've got everything sorted out. Ewa Sawicka's a good head of department. She's very fair to the people who work under her.

NINA That's important when you're a boss, isn't it?

ANNA It certainly is. Now, who else should I tell you about? Oh, yes. Monika Olszewska.

NINA Monika ...?

ANNA Olszewska. Don't worry. You'll get used to people's names quite quickly.

NINA I hope so. OK, Ms Olszweska, where does she fit in?

ANNA She's the production manager – in charge of Production. She looks after everything to do with production – quality control, after-sales service – and, on top of that, she takes care of the packaging and storage departments, too.

NINA Wow. She must be busy.

ANNA Yes, I suppose so. You won't have much to do with her. Lucjan Jakubiak is directly responsible to her.

NINA Lucjan's the production supervisor, right?

ANNA That's right, Nina, Lucjan's in charge of the production line. You and the other trainees all report to him.

## 5 Unit 2, Exercises 7 & 8

DAMEK Sorry Viktor, we'll have to leave it for the moment. I have another appointment so I don't have time to take you into the mechatronics workshop.

VIKTOR Oh, right.

DAMEK Tell you what. It's almost lunchtime. Why don't you go across to the canteen and have some lunch? We can meet again here at secondary production 2 at one o'clock. Then I'll take you over to the workshop and introduce you to Vavrin Dvorak. He'll be your training supervisor while you work here.

VIKTOR OK. Could you tell me how to get to the canteen?

DAMEK Yes, of course. Normally, I'd take you there myself, but I'm in such a hurry ... it's easy enough to find it, though. I'll tell you how to get there. Go out of the door here and turn left. Then walk past the loading bays and then take the first on the left. Go straight on, then take the second on the right ...

VIKTOR Just a moment, please. So, I go out here and turn left, then past the loading bays and left again. Then I take the first on the right ...

DAMEK The *second* on the right.

VIKTOR Oh, OK, *second* on the right.

DAMEK Then take the first left between Human Resources and the sports club. Go right after the sports club and you'll see the entrance to the canteen on your left. It's opposite the door to the sports club.

VIKTOR OK ... so that's left out the door here, past the loading bays, then left again, then second on the right and the next left. Then I have to turn right between Human Resources and the sports club and the door's opposite the door to the sports club.

DAMEK Exactly. You can't miss it. Enjoy your lunch. See you back here at one o'clock.

VIKTOR Yes. OK, one o'clock it is. ... Now, what did he say? Left, loading bay, left again, then ... ??? Oh, no! I'm lost.

## 6 Unit 2, Exercises 10 & 12

DAMEK Well, Viktor, here's the mechatronics workshop and this is Vavrin Dvorak, your training supervisor.

VAVRIN Ah, the new apprentice from Germany. Nice to meet you, Viktor.

VIKTOR Nice to meet you, too.

DAMEK OK. Can I leave you to it?

VAVRIN Yes, thanks, Damek. See you later. OK, Viktor, I'll start by introducing you to another apprentice, Jarek. He's coming to the end of his apprenticeship so it makes sense if he looks after you while you're here. If you have any problems, he'll help you to sort them out.

VIKTOR Right.

VAVRIN Well, to start with, let's have a look around the workshop. You'll be spending a lot of time here in the next couple of months so it's important that you know where everything is. I'll ask Jarek to show you the tools we use.

VIKTOR That's good. I'll get to know what they're all called in English.

VAVRIN Exactly. Once you know what all the tools are, you can do a routine maintenance check on one of our packaging machines. OK? After I've shown you round today, you can read the job specification for the maintenance check and then first thing tomorrow morning you can carry out the check.

VIKTOR Er, do you mean I have to do it alone?

VAVRIN No, no. Don't worry, Jarek will do it with you. He's done the job several times before so he's an expert at it.

VIKTOR Oh, fine.

VAVRIN Right. I'll just go and see if Jarek's here, then I can introduce you and we can get started.

## 8 Unit 3, Exercise 5

HELENA GBL Electronics, good morning.

JAN Oh, good morning, Helena. It's Jan here. Do you have a minute?

HELENA Oh, hi. Jan. How are you?

JAN Fine, thanks. You too?

HELENA Yes, thanks. So what can I do for you?
JAN Do you remember the multimeter I bought from you a couple of months ago?
HELENA Yes ... Everything's OK, I hope.
JAN Well, no, not really.
HELENA Oh dear. What's the problem?
JAN Well, there's something wrong with it. It's not giving correct readings.
HELENA Have you checked the battery?
JAN The battery. Oh, right.

## 9 Unit 3, Exercise 13

LUKAS Can you have a look at this with me, please, Magda?
MAGDA Sure. What's the problem?
LUKAS I'm trying to diagnose a fault. Look, the curve shoots up suddenly.
MAGDA Yes. That means that the diode breaks down.
LUKAS Oh, I see. Well, it breaks down at 0.5 in the positive direction. That's OK, isn't it?
MAGDA Yes, that's fine.
LUKAS But then it breaks down in the negative direction at -100V.
MAGDA That's not good. -400V is normal for this diode. Let's have a look at one which is working ...

## 11 Unit 4, Exercise 9

L. BLUM Good morning, ladies and gentlemen. Welcome to CCS, Germany. My name is Lance Blum. I'm the CCS representative here in Berlin. As you know, CCS Germany is a subsidiary of CCS USA. We manufacture all kinds of components and distribute them all over the world. Now, I know that you've come along here this morning to have a look around our factory but, before we start, I'd like to tell you about our components. I think this will help you get an overall picture of things before we start our tour. You're all mechatronics apprentices, right? Well, here's your chance to find out all you've ever wanted to know about components. I have an overview here of everything that we produce and I'll go through it, looking at the components one by one. Please ask any questions. I don't mind interruptions. So, to begin with, I'd like to start by talking about transistors. Now, who can tell me what a transistor is?

## 13 Unit 5, Exercise 13

MS WELLER PP Services, Weller. Guten Tag.
ALAIN Hello. This is Alain Dufour. I'm calling from Colmar. Is that Ms Weller?
MS WELLER Yes, it is. I'm very glad you've phoned, Mr Dufour. There's a bit of trouble ...
ALAIN A *bit* of trouble? I'd say that there's more than a *bit* of trouble. I've been waiting here for hours for your technicians to install the pneumatics for a new machine. Do you know if they'll be here soon?
MS WELLER I hope so, Mr Dufour. I'm afraid there's been an accident on the motorway and the technicians are caught up in a tailback. They tried to call you, but they couldn't get through.
ALAIN Well, I can't understand that. My mobile's on. Anyway – how much longer do you think they'll be?
MS WELLER I can't say exactly. Maybe two to three hours.
ALAIN Two to three hours? They're already two hours late. If they don't get the compressed air supply set up in time, the people from Switzerland won't be able to

install the machine according to the schedule. I'll be way behind with my orders ...

MS WELLER I'm sure that they'll ...

ALAIN Listen, Ms Weller. I know that this isn't your fault, but I'm very unhappy about this delay. I am really very angry about this. I made it absolutely clear that we need the machine to be up and running in the next two days. We'll have enormous production problems here if it isn't.

MS WELLER I can understand that you're upset. And I really must apologize for this. However, I don't think you need to worry. If the team arrives within the next three hours, they'll still be able to get everything installed in time. They'll work through the night if necessary.

ALAIN Well, OK. But if they don't manage to get the work done in time, I'll have to think seriously about whether we'll use your firm's services again.

MS WELLER I'm sure it won't come to that, Mr Dufour. We've always done a good job in the past.

## 14 Unit 5, Exercise 15

MS ROLFF RR Transportmaschinen, Rolff. Guten Morgen.

ALAIN Oh. Hello! Erm, do you speak English?

MS ROLFF Yes, I do. How can I help you?

ALAIN This is Alain Dufour calling from Orbis in Colmar, France.

MS ROLFF Good morning, Mr Dufour.

ALAIN Sorry, who am I speaking to, please?

MS ROLFF Janine Rolff. In the sales department.

ALAIN Ah. Ms Rolff. Is Mr Kahn in the office, please?

MS ROLFF No, I'm afraid he's not. He's in a meeting. Can I take a message?

ALAIN Yes, please. Could you tell him that there are some pages missing in the manual which he sent along with the RR45?

MS ROLFF Oh, dear. I'm sorry to hear that. I'll tell Mr Kahn and I'll arrange for a new manual to be sent out to you straight away.

ALAIN Thanks. But is there any chance you could fax or email me the missing pages first? We can't use the machine, you see.

MS ROLFF Oh, of course. Which pages are missing exactly?

ALAIN It's pages 18 to 21.

MS ROLFF Pages 18 to 21. Right. If you give me your fax number, I'll fax those pages to you straight away.

ALAIN Thank you very much, Ms Rolff. That's what I call service! My fax number is ...

## 16 Unit 6, Exercise 16

ALAIN So, here's the tube production machine. And as I mentioned, we may have a problem with the hydraulic system.

ISABELL OK. Let's just take a look at it. I'll check the power pack first of all. Right, here it is.

ALAIN What do you think the problem is?

ISABELL It's hard to say until I've gone over the main components.

ALAIN Hm. What is all this stuff anyway?

ISABELL It looks complicated but it's really quite straightforward. I'll talk you through it if you like.

ALAIN Good. Maybe I can repair it myself next time.

ISABELL Maybe. Look here. The large cylinder on the right is the accumulator. Now look to the right – the small tube with the grey cap is the theromostatic valve.

ALAIN Thermostatic valve. OK. And what's that?
ISABELL This here? The flat component directly in front of the accumulator?
ALAIN Yes, I can't see it very clearly behind all those pipes, but it looks like a metal plate lying on top of the tank, doesn't it?
ISABELL That's right, that's the heat exchanger. Now, if you look to the left of the heat exchanger, on the far left-hand side, you'll see a tall black component.
ALAIN You mean that component with all the cables coming out of it?
ISABELL Exactly. That's the valve block. At the moment we're using four valves.
ALAIN And that's where the cables are attached?
ISABELL That's right. But as you can see, there is actually room for more valves.
ALAIN I see. And at the other end, the cables are connected to that box which is hanging down in front of the accumulator. What is that?
ISABELL That's the sensor/actuator box.
ALAIN The what?
ISABELL The sensor/actuator box. It distributes the signals and triggers the valves.
ALAIN ... triggers the valves. OK. And what about this here, the big blue drum behind the valve block? Looks like an electric motor.
ISABELL Correct, that's the motor. And down at the bottom – that's the tank. Oh, and this indicator on the right of the tank is the oil level indicator, it's slightly hidden by the bar across the front of the tank, you can just see a black switch in the corner.
ISABELL Well, now you've explained it, it all seems pretty straightforward. Just as you said.
ALAIN Do you want to have a go at fixing it yourself then?
ALAIN Oh, no. I'll leave that up to you.
ISABELL I thought you'd say that!

## 18 Unit 7, Exercise 9

GORDON Right ... and that brings me to the end of my presentation. I can see that one or two of you have been waiting patiently to ask me questions about what I've said and now's your chance. Oh, and could you tell me your name before you ask your questions, please? I like to know who I'm talking to. Yes ... would you like to start?
JENNY Yes, please. My name's Jenny Sager. I was a bit confused about something you said. Can we look at inputs and outputs again? Did you say that inputs are the information which the sensors send to the CPU? And then the CPU uses this information to control the machine?
GORDON That's right Jenny. You've understood it perfectly. I couldn't have put it better myself. Anyone else? Yes ... the young man over there with the red jumper.
BERNDT I'm Berndt Anders. Could you clarify what you said about connecting PCs, please? You talked about PC/PPI cables, communication processor cards and multipoint interface cables. Do you mean that we need all of these before we can use a PC with the S7-200?
GORDON No, not quite. You need either a PC/PPI cable or a communication processor card and a multipoint interface cable. OK? Does that answer your question?
BERNDT Yes it does, thanks.
GORDON Good. Now. I think we have time to fit in one more question. Yes ... over there.
STEFAN Hi. I'm Stefan Braun. I just want to clear up a point about installation. You said that we need to leave at least 25 mm around the CPU if we're installing extra modules. If I understood you correctly, you mean that to engage and disengage the bus expansion connector we need to leave a space of at least 25 mm. Is that right?
GORDON Yes, Stefan. That's spot on. That's exactly what I meant. Right now, let's stop there because it's almost time for lunch. If there are any more questions, we can talk about them over lunch. And don't forget that there'll be more time for questions during this afternoon's practical hands-on session.

## 19 Unit 8, Exercise 13

AGNESE So, Hannah. Why did you decide to become a mechatronics technician?

HANNAH Well, I studied physics for my Abitur – that's our final school examination – and I really enjoyed it. I wanted to do something practical when I left school, and mechatronics seemed like the most modern and exciting choice.

AGNESE Good. I see from your CV that you've had some work experience. Could you tell me about that, please?

HANNAH Of course. When I left school, I went to work in a small, local engineering company. I really enjoyed that. I think that's what finally helped me make up my mind to become a mechatronics technician as a matter of fact.

AGNESE That's nice. ... I see from your letter of application that you're about to finish your apprenticeship.

HANNAH That's right. I'll be finished in two weeks' time.

AGNESE And you're looking forward to getting your first job. Why are you interested in coming to work in Latvia, by the way?

HANNAH Well, I think it's really important to get experience of working outside Germany. I need to learn about how people work in other countries. And I'd like to improve my English which I'd have to use every day if I came to work here.

AGNESE That's true. Now, you've read all the details about the job here and the type of person we're looking for and you know that sensors will be the key focus of the job. What have you learned about sensors during your training?

HANNAH Hm, in the first year we looked at all the different types of sensors and what you can do with them.

AGNESE And what about troubleshooting sensor systems problems? What experience have you gained in this area in the last three years?

HANNAH Well, during the second year of my apprenticeship I learned how to analyze problems with sensors. It's really important to find out quickly what's causing a sensor fault so that there's not too much downtime in production.

AGNESE That's correct. And we also think that it's important to monitor sensor performance on an ongoing basis to avoid problems occurring. What have you learned about that since you started your apprenticeship?

HANNAH Oh, I've learned about monitoring and maintenance of all sorts of systems, including sensors. I think I've gained quite a bit of experience in this since I began my training.

AGNESE Good. Well, you're going to be doing a test this afternoon, so you'll get a chance to show us what you know. In the meantime, I would like to finish by asking you two more questions. What are your strengths and weaknesses?

HANNAH Hm. My strengths – I'm a fast learner. I grasp things quickly and I'm keen to learn. Erm ... my weaknesses – well, perhaps sometimes I'm a bit too spontaneous – I want to get on with work before I've planned it properly.

AGNESE Oh. And, finally, could you tell me in one sentence why would like this job?

HANNAH Because I think it'll be a really challenging opportunity for me to work in a foreign company and get my career off to a good start.

AGNESE Fine, Hannah. Thanks. Well, I think that brings us to the test.

# Unit word list

Dieses Wörterverzeichnis enthält alle Wörter, die nicht in der Liste des Grundwortschatzes enthalten sind, d.h. die nicht als bekannt vorausgesetzt werden. t = das Wort befindet sich in den *Transcripts* (Hörverständnistexten). p = das Wort befindet sich in den *Pairwork Files*.

## Unit 1

| | | |
|---|---|---|
| 6 | **warm-up** ['wɔ:m ʌp] | Vorbereitung, Einstimmung, Einstieg |
| | **get to know** [get tə 'nəʊ] | kennen lernen |
| | **solder** ['səʊldə] | löten |
| | **repair** [rɪ'peə] | reparieren |
| | **introduction** [ˌɪntrə'dʌkʃn] | Einführung |
| | **mechatronics** [ˌmekə'trɒnɪks] | Mechatronik |
| | **mechatronics technician** [mekəˌtrɒnɪks tek'nɪʃn] | Mechatroniker/in |
| | **engineering** [ˌendʒɪ'nɪərɪŋ] | Maschinenbau |
| | **subsidiary** [səb'sɪdiəri] | Tochter(gesellschaft) |
| | **receptionist** [rɪ'sepʃənɪst] | Empfang(smitarbeiter/in) |
| | **appointment** [ə'pɔɪntmənt] | Termin, Verabredung |
| | **HR (Human Resources) Manager** [ˌeɪtʃ 'ɑ:, ˌhju:mən rɪ'sɔ:sɪz mænɪdʒə] | Personalleiter/in, -chef/in |
| | **give sb a call** [gɪv ə 'kɔ:l] | jdn anrufen |
| | **over there** [ˌəʊvə 'ðeə] | dort (drüben) |
| 7 | **How do you do?** [ˌhaʊ dju 'du:] | Guten Tag/Morgen/Abend! |
| | **introduce** [ˌɪntrə'dju:s] | (sich) vorstellen |
| | **training manager** ['treɪnɪŋ mænɪdʒə] | Ausbildungsleiter/in |
| | **take sb through sth** [ˌteɪk 'θru:] | jdn durch etw führen |
| | **employee** [ɪm'plɔɪi:] | Angestellte/r |
| | **induction programme** [ɪn'dʌkʃn prəʊgræm] | Arbeitseinführungsprogramm |
| | **be based** [bi 'beɪst] | liegen, angesiedelt sein |
| | **headquarters** [ˌhed'kwɔ:təz] | Hauptsitz, Zentrale |
| | **shopfloor** [ˌʃɒp'flɔ:] | Produktion(sbereich) |
| | **production supervisor** [prəˌdʌkʃn 'su:pəvaɪzə] | Produktionsleiter/in |
| | **responsible for** [rɪ'spɒnsəbl] | verantwortlich/zuständig für |
| | **trainee** [treɪ'ni:] | Auszubildende/r, Praktikant/in |
| | **workshop** ['wɜ:kʃɒp] | Werkstatt, Werkshalle |
| | **formal** ['fɔ:ml] | formell, förmlich |
| | **join** [dʒɔɪn] | sich anschließen, nachkommen |
| | **lathe** [leɪð] | Drehbank |
| | **production line robot** [prəˌdʌkʃn laɪn 'rəʊbɒt] | Fertigungsbandroboter |
| | **CNC (Computerized Numerical Control)** [ˌsi: en 'si:, kəmˌpju:təraɪzd nju:ˌmerɪkl kən'trəʊl] | Computerisierte Numerische Steuerung |
| | **be familiar with** [bi fə'mɪliə wɪð] | vertraut sein mit, kennen |
| 8 | **statement** ['steɪtmənt] | Aussage, Feststellung |
| | **staff** [stɑ:f] | Angestellte, Personal |
| | **branch** [brɑ:ntʃ] | Filiale, Zweigstelle, Niederlassung |
| | **apprentice** [ə'prentɪs] | Lehrling, Auszubildende/r |
| | **look after** [lʊk 'ɑ:ftə] | sich kümmern um, betreuen |
| | **machinery** [mə'ʃi:nəri] | Maschinen(park), Geräte |
| | **machine tool** [mə'ʃi:n tu:l] | Werkzeugmaschine |
| | **conveyor belt** [kən'veɪə belt] | Fließ-, Förderband |
| | **electrical discharge machine** [ɪˌlektrɪkl 'dɪstʃɑ:dʒ məʃi:n] | Funkenerosionsmaschine |
| | **laser welder** ['leɪzə weldə] | Laserschweißgerät |
| | **milling machine** ['mɪlɪŋ məʃi:n] | Fräsautomat |
| | **pulley** ['pʊli] | Flaschenzug, Rolle |
| | **moving** ['mu:vɪŋ] | laufend |
| | **band** [bænd] | Band |
| | **workpiece** ['wɜ:kpi:s] | Werkstück |
| | **melt** [melt] | schmelzen |
| | **shape** [ʃeɪp] | formen |
| | **basically** ['beɪsɪkli] | im Grunde, eigentlich |
| | **wheel** [wi:l] | Rad |
| | **rope** [rəʊp] | Seil, Tau |
| | **lift** [lɪft] | (an-, hoch)heben |
| | **load** [ləʊd] | Last |
| | **manually-controlled** ['mænjuəli kəntrəʊld] | von Hand bedient |
| | **drill** [drɪl] | (durch)bohren |
| | **pick** [pɪk] | auswählen, mitnehmen |
| | **joint** [dʒɔɪnt] | Gelenk |
| | **spin** [spɪn] | (schnell) drehen |
| 10 | **noon** [nu:n] | Mittag |
| | **order** ['ɔ:də] | Reihenfolge |
| | **small talk** ['smɔ:l tɔ:k] | höfliche Konversation, Smalltalk |
| | **response** [rɪ'spɒns] | Reaktion, Antwort |
| | **time of year** [ˌtaɪm əv 'jɪə] | Jahreszeit |
| | **comfortable** ['kʌmftəbl] | bequem, behaglich |
| | **be delayed** [bi dɪ'leɪd] | Verspätung haben |
| | **in the end** [ɪn ði 'end] | zum Schluss, schließlich |

| | | |
|---|---|---|
| | **look forward to** [ˌlʊk 'fɔːwəd tə] | sich freuen auf |
| | **get started** [get 'stɑːtɪd] | anfangen |
| 11 | **Anglo-American** [ˌæŋgləʊ ə'merɪkən] | angloamerikanisch |
| | **common** ['kɒmən] | normal, üblich |
| | **shake hands** [ʃeɪk 'hændz] | sich die Hand geben |
| | **wait and see** [ˌweɪt ən 'siː] | abwarten |
| | **introduce oneself** [ˌɪntrə'djuːs wʌnself] | sich vorstellen |
| | **over to sb** ['əʊvə tə] | jd ist dran |
| | **practise** ['præktɪs] | (ein)üben, trainieren |
| | **structure** ['strʌktʃə] | Struktur, Aufbau |
| | **production manager** [prə'dʌkʃn mænɪdʒə] | Produktionsleiter/in |
| | **dispatch supervisor** [dɪˌspætʃ 'suːpəvaɪzə] | Versandleiter/in |
| | **marketing manager** ['mɑːkɪtɪŋ mænɪdʒə] | Marketingleiter/in |
| | **head of accounts** [ˌhed əv əˌkaʊnts] | Leiter/in Buchhaltung |
| 11t | **section** ['sekʃn] | Abteilung |
| | **take a look at** [teɪk ə 'lʊk ət] | (sich) ansehen, betrachten |
| | **talk sb through sth** [ˌtɔːk 'θruː] | jdm etw erklären |
| | **first of all** [fɜːst əv 'ɔːl] | zu(aller)erst, zunächst |
| | **for short** [fə 'ʃɔːt] | abgekürzt |
| | **personnel** [ˌpɜːsə'nel] | Personal, Mitarbeiter |
| | **pay** [peɪ] | Lohn, Gehalt, Entlohnung, Bezahlung |
| | **training** ['treɪnɪŋ] | Ausbildung, Schulung, Training |
| | **responsible to sb** [rɪ'spɒnsəbl tə] | jdm unterstellt |
| | **level** ['levl] | Ebene, Stufe |
| | **in charge of** [ɪn 'tʃɑːdʒ əv] | verantwortlich für |
| | **accounts** [ə'kaʊnts] | Buchhaltung, Rechnungswesen |
| | **marketing** ['mɑːkɪtɪŋ] | Marketing, Werbung, Vertrieb |
| | **sales** [seɪlz] | Vertrieb, Verkauf |
| | **advertising** ['ædvətaɪzɪŋ] | Werbung, Reklame |
| | **deal with** ['diːl wɪð] | sich befassen mit, bearbeiten, erledigen |
| | **complaint** [kəm'pleɪnt] | Beschwerde, Beanstandung, Reklamation |
| | **imagine** [ɪ'mædʒɪn] | sich vorstellen |
| | **especially** [ɪ'speʃəli] | besonders |
| | **polite** [pə'laɪt] | höflich |
| | **get on with** [ˌget 'ɒn wɪð] | auskommen mit, sich verstehen mit |
| | **dispatch** [dɪ'spætʃ] | Versand, Auslieferung |
| | **hectic** ['hektɪk] | hektisch |
| | **down there** ['daʊn ðeə] | dort unten |
| | **storm** [stɔːm] | stürmen |
| | **upstairs** [ˌʌp'steəz] | die Treppe hinauf, (nach) oben |
| | **complain** [kəm'pleɪn] | (sich) beschweren |
| | **go over** [ˌgəʊ 'əʊvə] | durchgehen |
| | **sort out** [ˌsɔːt 'aʊt] | in Ordnung bringen, klären, lösen |
| | **head of department** [ˌhed əv dɪ'pɑːtmənt] | Abteilungsleiter/in |
| | **else** [els] | sonst (noch) |
| | **get used to** [get 'juːst tə] | sich gewöhnen an |
| | **fit in** [ˌfɪt 'ɪn] | (sich) einfügen (in), gehören |
| | **to do with** [tə 'duː wɪð] | die zu tun haben mit |
| | **quality control** ['kwɒləti kəntrəʊl] | Qualitätskontrolle |
| | **after-sales service** ['ɑːftə seɪlz sɜːvɪs] | Kundendienst, Wartung |
| | **on top of that** [ɒn 'tɒp əv ðət] | darüber hinaus |
| | **take care of** [teɪk 'keər əv] | sich kümmern um |
| | **packaging** ['pækɪdʒɪŋ] | Verpackung |
| | **storage** ['stɔːrɪdʒ] | Lager(haltung) |
| | **department** [dɪ'pɑːtmənt] | Abteilung |
| | **suppose** [sə'pəʊz] | glauben, annehmen |
| | **production line** [prə'dʌkʃn laɪn] | Fertigung(sband), Produktion |
| 12 | **organization chart** [ˌɔːgənaɪ'zeɪʃn tʃɑːt] | Organigramm |
| | **chart** [tʃɑːt] | Diagramm, Tabelle |
| | **resources** [rɪ'sɔːsɪz] | Mittel, Ressourcen |
| | **relationship** [rɪ'leɪʃnʃɪp] | Beziehung, Verhältnis |
| | **report to** [rɪ'pɔːt tə] | unterstehen, unterstellt sein |
| | **responsibility (for)** [rɪˌspɒnsə'bɪləti fə] | Verantwortung (für) |
| | **technician** [tek'nɪʃn] | Techniker/in |
| | **supervisor** ['suːpəvaɪzə] | Leiter/in |
| | **work for** ['wɜːk fə] | arbeiten für |
| | **strict** [strɪkt] | streng |
| | **overall** [ˌəʊvər'ɔːl] | (ins)gesamt, Gesamt-, umfassend |
| | **take turns** [teɪk 'tɜːnz] | (sich) abwechseln |
| | **phrase** [freɪz] | (Rede-)Wendung, Ausdruck |
| 13 | **bracket** ['brækɪt] | Klammer |
| | **report** [rɪ'pɔːt] | berichten |
| | **design** [dɪ'zaɪn] | konstruieren, entwerfen, entwickeln |
| | **sales department** ['seɪlz dɪpɑːtmənt] | Verkauf(sabteilung) |
| | **file** [faɪl] | ablegen |
| | **handle** ['hændl] | erledigen, bearbeiten |
| | **organize** ['ɔːgənaɪz] | organisieren |
| | **recruit** [rɪ'kruːt] | einstellen |
| | **payment** ['peɪmənt] | Zahlung, Auszahlung |
| | **salary** ['sæləri] | Gehalt, Lohn |
| | **benefit** ['benɪfɪt] | zusätzliche Leistung |
| | **insurance** [ɪn'ʃʊərəns] | Versicherung |
| | **training session** ['treɪnɪŋ seʃn] | Trainingsstunde |
| | **currently** ['kʌrəntli] | momentan, zurzeit |
| | **including** [ɪn'kluːdɪŋ] | einschließlich, inbegriffen, darunter |

| | | |
|---|---|---|
| **14** | **health and safety** [ˌhelθ ən 'seɪfti] | Arbeitsschutz |
| | **warning** ['wɔ:nɪŋ] | Warnung, Hinweis, Warn- |
| | **sign** [saɪn] | Schild |
| | **meaning** ['mi:nɪŋ] | Bedeutung |
| | **hard hat** [ˌhɑ:d 'hæt] | Schutzhelm |
| | **area** ['eəriə] | Bereich |
| | **first aid** [ˌfɜ:st 'eɪd] | erste Hilfe |
| | **fire extinguisher** ['faɪər ɪkstɪŋgwɪʃə] | Feuerlöscher |
| | **voltage** ['vəʊltɪdʒ] | Spannung |
| | **emergency** [ɪ'mɜ:dʒənsi] | Notfall |
| | **safety glove** ['seɪfti glʌv] | Schutzhandschuh |
| | **ear protection** [ˌɪə prə'tekʃn] | Gehörschutz |
| | **goggles** ['gɒglz] | Schutzbrille |
| | **safety boot** ['seɪfti bu:t] | Sicherheitsschuh |
| | **guideline** ['gaɪdlaɪn] | Richtlinie, Hinweis |
| | **access** ['ækses] | Zugang |
| | **attention** [ə'tenʃn] | Aufmerksamkeit |
| | **designated** ['dezɪgneɪtɪd] | bestimmt, gekennzeichnet |
| | **distract** [dɪ'strækt] | ablenken |
| | **hazardous** ['hæzədəs] | risikoreich, gefährlich |
| | **instruct** [ɪn'strʌkt] | anweisen, beauftragen |
| | **loose** [lu:s] | weit (geschnitten), lose |
| | **operate** ['ɒpəreɪt] | bedienen, betätigen |
| | **spill** [spɪl] | ausschütten, verschütten |
| | **serious(ly)** ['sɪəriəs, -li] | schwer, ernst(haft) |
| | **jewellery** ['dʒu:əlri] | Schmuck(stücke) |
| | **catch on** [ˌkætʃ 'ɒn] | erfasst werden von, sich verfangen in |
| | **equipment** [ɪ'kwɪpmənt] | Maschinen, Geräte |
| | **cause** [kɔ:z] | verursachen |
| | **injury** ['ɪndʒəri] | Verletzung |
| | **injure** ['ɪndʒə] | verletzen |
| | **necessary** ['nesəsəri] | nötig, erforderlich |
| | **shut down** [ˌʃʌt 'daʊn] | aus-, abschalten |
| | **carry out** [ˌkæri 'aʊt] | aus-, durchführen |
| | **maintenance** ['meɪntənəns] | Wartung, Pflege |
| | **block** [blɒk] | blockieren, versperren |
| | **properly** ['prɒpəli] | richtig, korrekt |
| | **authorize** ['ɔ:θəraɪz] | autorisieren, genehmigen |
| | **passage** ['pæsɪdʒ] | Gang, Durchgang |
| | **shortcut** ['ʃɔ:tkʌt] | Abkürzung |
| | **clean up** [ˌkli:n 'ʌp] | säubern, aufwischen |
| | **liquid** ['lɪkwɪd] | Flüssigkeit |
| | **oil** [ɔɪl] | Öl |
| | **grease** [gri:s] | Schmiere, Fett |
| | **mention** ['menʃn] | erwähnen |
| **15** | **in the event of** [ɪn ði ɪ'vent əv] | im Falle eines |
| | **accident** ['æksɪdənt] | Unfall |
| | **instruction** [ɪn'strʌkʃn] | Anweisung, Hinweis |
| | **assistance** [ə'sɪstəns] | Hilfe(stellung) |
| | **major** ['meɪdʒə] | groß, größer |
| | **body trauma** [ˌbɒdi 'trɔ:mə] | schwere Körperverletzung |
| | **ambulance** ['æmbjələns] | Krankenwagen |
| | **mark** [mɑ:k] | kennzeichnen |

| | |
|---|---|
| **acquaint oneself with** [ə'kweɪnt wʌnself wɪð] | sich vertraut machen mit |
| **contents** [kən'tents] | Inhalt |
| **contain** [kən'teɪn] | enthalten |
| **sterile** ['steraɪl] | steril |
| **bandage** ['bændɪdʒ] | Verband, Binde |
| **antiseptic** [ˌænti'septɪk] | antiseptisch |
| **ointment** ['ɔɪntmənt] | Salbe |
| **eye pad** ['aɪ pæd] | Augenverband |
| **tape** [teɪp] | Wund-, Heftpflaster |
| **eye wash** ['aɪ wɒʃ] | Augenspülung |
| **safety pin** ['seɪfti pɪn] | Sicherheitsnadel |
| **gauze pad** ['gɔ:z pæd] | Mullverband |
| **severe** [sɪ'vɪə] | schwer, stark |
| **bleeding** ['bli:dɪŋ] | Blutung |
| **procedure** [prə'si:dʒə] | Vorgehensweise, Maßnahme |
| **wound** [wu:nd] | Wunde |
| **apply sth to sth** [ə'plaɪ tə] | etw auf etw ausüben, anwenden |
| **pressure** ['preʃə] | Druck |
| **persist** [pə'sɪst] | andauern |
| **major artery** [ˌmeɪdʒər 'ɑ:təri] | Hauptarterie |
| **supply** [sə'plaɪ] | versorgen |
| **burn** [bɜ:n] | Verbrennung |
| **minor** ['maɪnə] | gering(fügig) |
| **skin** [skɪn] | Haut |
| **unbroken** [ʌn'brəʊkən] | nicht verletzt |
| **cover** ['kʌvə] | (be-, ab)decken |
| **seek** [si:k] | (auf)suchen |
| **medical** ['medɪkl] | medizinisch, ärztlich |
| **breathe** [bri:ð] | atmen |
| **administer** [əd'mɪnɪstə] | (durch)führen, anwenden |
| **mouth-to-mouth resuscitation** [maʊθ tə ˌmaʊθ rɪˌsʌsɪ'teɪʃn] | Mund-zu-Mund-Beatmung |
| **straight away** [streɪt ə'weɪ] | sofort, gleich, auf der Stelle |
| **remove** [rɪ'mu:v] | entfernen, beseitigen |
| **smoulder** ['sməʊldə] | schwelen, glimmen |
| **moist** [mɔɪst] | feucht, nass |
| **blister** ['blɪstə] | Blase |
| **shock** [ʃɒk] | Schock, Schlag |
| **turn off** [ˌtɜ:n 'ɒf] | aus-, abschalten |
| **source** [sɔ:s] | Quelle |
| **electricity** [ɪˌlek'trɪsəti] | Elektrizität, Strom |
| **blanket** ['blæŋkɪt] | (Woll-)Decke |
| **allow** [ə'laʊ] | gewähren, ermöglichen, (zu)lassen |
| **recovery** [rɪ'kʌvəri] | Erholung |
| **pulse** [pʌls] | Puls |
| **breathing** ['bri:ðɪŋ] | Atmung |
| **cardiopulmonary resuscitation** [kɑ:diəʊˌpʌlmənəri rɪˌsʌsɪ'teɪʃn] | Herzmassage, künstliche Beatmung |
| **treat** [tri:t] | behandeln |
| **as instructed** [əz ɪn'strʌktɪd] | wie angegeben |
| **translate** [træns'leɪt] | übersetzen, übertragen |

| | | |
|---|---|---|
| | **item** ['aɪtəm] | Artikel, Gegenstand |
| | **dust** [dʌst] | Staub |
| | **irritant** ['ɪrɪtənt] | Reizstoff |
| | **fasten** ['fɑ:sn] | befestigen, anlegen |
| | **keep in place** [ki:p ɪn 'pleɪs] | befestigen |
| | **protect** [prə'tekt] | schützen |
| | **dress** [dres] | verbinden |
| | **irritate** ['ɪrɪteɪt] | reizen |
| | **life-saving** ['laɪf seɪvɪŋ] | Lebensrettungs- |
| | **treatment** ['tri:tmənt] | (medizinische) Behandlung |
| | **expression** [ɪk'spreʃn] | (Gesichts-)Ausdruck |
| | **mediation** [ˌmi:di'eɪʃn] | Vermittlung |
| | **workplace** ['wɜ:kpleɪs] | Arbeitsplatz |
| | **rule** [ru:l] | Vorschrift, Regel |

## Unit 2

| | | |
|---|---|---|
| 16 | **apprenticeship** [ə'prentɪʃɪp] | Lehre, Ausbildung |
| | **site** [saɪt] | (Arbeits-)Stelle, (Firmen-)Gelände, Anlage |
| | **production engineer** [prəˌdʌkʃn ˌendʒɪ'nɪə] | Betriebsingenieur/in |
| | **sound** [saʊnd] | (sich) anhören, klingen |
| | **admin (administration)** ['ædmɪn, ədˌmɪnɪ'streɪʃn] | Verwaltung |
| | **finance** ['faɪnæns] | Finanz(en), Buchhaltung |
| 17 | **production hall** [prə'dʌkʃn hɔ:l] | Produktionshalle |
| | **delivery bay** [dɪ'lɪvəri beɪ] | Lieferzone |
| | **raw material** [ˌrɔ: mə'tɪəriəl] | Rohstoff(e) |
| | **row** [rəʊ] | Reihe |
| | **R&D (research and development)** [rɪˌsɜ:tʃ ən dɪ'veləpmənt] | Forschung und Entwicklung |
| | **at the far end** [ət ðə fɑ:r 'end] | am anderen Ende |
| | **warehouse** ['weəhaʊs] | Lager(halle, -haus) |
| | **training centre** ['treɪnɪŋ sentə] | Ausbildungszentrum |
| | **secondary production** [ˌsekəndri prə'dʌkʃn] | verarbeitende Produktion |
| | **primary production** [ˌpraɪməri prə'dʌkʃn] | Grundstoffproduktion |
| | **process** ['prəʊses] | (weiter)verarbeiten, bearbeiten |
| | **pretty** ['prɪti] | hier: ziemlich |
| | **impressive** [ɪm'presɪv] | beeindruckend, eindrucksvoll |
| | **full automation** [fʊl ˌɔ:tə'meɪʃn] | volle Automation |
| | **robotics** [rəʊ'bɒtɪks] | Robotik, Automatisierung |
| | **in action** [ɪn 'ækʃn] | in Aktion |
| | **package** ['pækɪdʒ] | verpacken |
| | **latest** ['leɪtɪst] | neueste |
| | **loading bay** ['ləʊdɪŋ beɪ] | Ladebucht, -zone |
| | **load onto** [ləʊd 'ɒntə] | aufladen |
| | **lorry** ['lɒri] | Lastwagen, Lkw |
| | **engineer** [ˌendʒɪ'nɪə] | Ingenieur/in, Techniker/in |
| | **next door to** [ˌnekst 'dɔ: tə] | neben |
| 18 | **efficient(ly)** [ɪ'fɪʃnt, -li] | effizient, (leistungs)fähig |
| | **site plan** ['saɪt plæn] | Lageplan |
| | **changing room** ['tʃeɪndʒɪŋ ru:m] | Umkleide(raum) |
| 19 | **administration block** [ədˌmɪnɪ'streɪʃn blɒk] | Verwaltung(sgebäude) |
| | **sequence** ['si:kwəns] | Reihe(nfolge) |
| | **basement** ['beɪsmənt] | Untergeschoss, Keller(geschoss) |
| | **develop** [dɪ'veləp] | entwickeln |
| | **technique** [tek'ni:k] | Methode, (Arbeits-)Technik |
| | **unprocessed** [ˌʌn'prəʊsest] | nicht bearbeitet |
| | **substance** ['sʌbstəns] | Substanz, Stoff |
| | **commercial** [kə'mɜ:ʃl] | Handels-, kommerziell |
| | **function** ['fʌŋkʃn] | Funktion |
| | **perform** [pə'fɔ:m] | durchführen |
| | **supplies** [sə'plaɪz] | Vorräte, Material |
| | **store** [stɔ:] | lagern, aufbewahren |
| | **point** [pɔɪnt] | Punkt, Stelle |
| | **on the other side of** [ɒn ði 'ʌðə saɪd əv] | gegenüber |
| | **location** [ləʊ'keɪʃn] | (Stand-)Ort, Lage, Position |
| | **as** [əz] | während |
| | **direction** [də'rekʃn] | Anweisung |
| 19t | **right now** [raɪt 'naʊ] | sofort |
| | **tell you what** ['tel ju wɒt] | hör mal, hören Sie mal |
| | **go across** [gəʊ ə'krɒs] | hinübergehen |
| | **get to** ['get tə] | gehen zu, kommen zu |
| | **be in a hurry** [bi ɪn ə 'hʌri] | es eilig haben |
| | **easy enough** ['i:zi ɪnʌf] | ganz leicht |
| | **straight on** [ˌstreɪt 'ɒn] | geradeaus |
| | **lost** [lɒst] | verloren |
| 20 | **rather (than)** ['rɑ:ðə ðən] | anstatt |
| | **however** [haʊ'evə] | doch, jedoch |
| | **senior** ['si:niə] | vorgesetzt, leitend, ranghöher, älter |
| | **member of staff** [ˌmembər əv 'stɑ:f] | Mitarbeiter/in |
| | **tell otherwise** [tel 'ʌðəwaɪz] | etw anderes sagen |
| 20p | **plc (public limited company)** [ˌpi: el 'si:] | AG |
| 20 | **happen** ['hæpən] | passieren, geschehen |
| | **demonstration** [ˌdemən'streɪʃn] | Demonstration, Vorführung |
| | **tool** [tu:l] | Werkzeug, Gerät |
| | **routine** [ru:'ti:n] | Routine(arbeit), routinemäßig |
| | **maintenance check** ['meɪntənəns tʃek] | Wartung(stest) |
| | **compare** [kəm'peə] | vergleichen |
| | **event** [ɪ'vent] | Ereignis |

| | | |
|---|---|---|
| 20t | **glad** [glæd] | froh |
| | **show sb around** [ˌʃəʊ əˈraʊnd] | jdn herumführen |
| | **a couple of** [ə ˈkʌpl əv] | ein paar |
| | **chance** [tʃɑːns] | Chance, Gelegenheit, Möglichkeit |
| | **make sure** [ˌmeɪk ˈʃʊə] | sicherstellen, sich vergewissern |
| | **finish off** [ˌfɪnɪʃ ˈɒf] | beenden |
| | **specification** [ˌspesɪfɪˈkeɪʃn] | Angabe, Aufstellung |
| | **first thing** [ˈfɜːst θɪŋ] | als Allererstes, früh am Morgen |
| 21 | **alternative** [ɔːlˈtɜːnətɪv] | Alternative, Auswahlantwort |
| | **by oneself** [baɪ wʌnˈself] | allein |
| | **layout** [ˈleɪaʊt] | Grundriss, Plan |
| | **in italics** [ɪn ɪˈtælɪks] | kursiv |
| | **appropriate** [əˈprəʊpriət] | angemessen, passend |
| | **start up** [ˌstɑːt ˈʌp] | starten |
| | **warning light** [ˈwɔːnɪŋ laɪt] | Warnlicht |
| | **flash** [flæʃ] | blinken, leuchten |
| | **service** [ˈsɜːvɪs] | Service, Wartung, Inspektion |
| | **excellent** [ˈeksələnt] | ausgezeichnet, hervorragend |
| | **install** [ɪnˈstɔːl] | installieren |
| | **right away** [raɪt əˈweɪ] | sofort, umgehend |
| 22 | **recording** [rɪˈkɔːdɪŋ] | Aufnahme, Aufzeichnung |
| | **identify** [aɪˈdentɪfaɪ] | identifizieren, feststellen |
| | **tense** [tens] | Zeit(form) |
| | **trade** [treɪd] | Handwerk, Beruf |
| | **power drill** [ˈpaʊə drɪl] | elektrische Bohrmaschine, Elektrobohrer |
| | **hacksaw** [ˈhæksɔː] | Bügelsäge |
| | **folding rule** [ˈfəʊldɪŋ ruːl] | Zollstock |
| | **needle nose pliers** [ˈniːdl nəʊz plaɪəz] | Spitzzange |
| | **side cutting pliers** [ˈsaɪd kʌtɪŋ plaɪəz] | Seitenschneider |
| | **soldering iron** [ˈsəʊldərɪŋ aɪən] | Lötkolben |
| | **round nose pliers** [ˈraʊnd nəʊz plaɪəz] | Rundzange |
| | **wire strippers** [ˈwaɪə strɪpəz] | Abisolierzange |
| | **spanner** [ˈspænə] | Schraubenschlüssel |
| | **hammer** [ˈhæmə] | Hammer |
| | **pump pliers** [ˈpʌmp plaɪəz] | Wasserpumpenzange |
| | **Phillips screwdriver** [ˌfɪlɪps ˈskruːdraɪvə] | Kreuzschlitzschraubendreher |
| | **(slot-head/flat-bladed) screwdriver** [ˌslɒt hed, ˌflæt bleɪdɪd ˈskruːdraɪvə] | (Schlitz-)Schraubenzieher, -dreher |
| | **Allen key** [ˈælən kiː] | Inbusschlüssel, Innensechskantschlüssel |
| | **vernier calliper** [ˈvɜːniə kælɪpə] | Messschieber |
| | **multimeter** [ˌmʌltiˈmiːtə] | Vielfachmessinstrument |
| 23 | **tighten** [ˈtaɪtn] | anziehen, festschrauben |
| | **screw** [skruː] | Schraube |
| | **a pair of pliers** [ə ˌpeər əv ˈplaɪəz] | eine Zange |
| | **grip** [grɪp] | festhalten, (er)greifen |
| | **loosen** [ˈluːsn] | lockern, lösen |
| | **measure** [ˈmeʒə] | messen |
| | **pick up** [ˌpɪk ˈʌp] | aufheben |
| | **strip** [strɪp] | abisolieren |
| | **bolt** [bəʊlt] | Bolzen, Schraube |
| | **component** [kəmˈpəʊnənt] | (Bau-)Element, (Bau-)Teil |
| | **electrical circuit** [ɪˌlektrɪkl ˈsɜːkɪt] | Stromkreis |
| | **hole** [həʊl] | Loch |
| | **nut** [nʌt] | Mutter |
| | **discuss** [dɪˈskʌs] | besprechen, diskutieren (über) |
| | **workbench** [ˈwɜːkbentʃ] | Werkbank |
| | **powered** [ˈpaʊəd] | elektrisch (angetrieben) |
| | **oscilloscope** [əˈsɪləskəʊp] | Oszilloskop |
| | **resistor** [rɪˈzɪstə] | Widerstand |
| | **LED** [ˌel iː ˈdiː] | LED-Anzeige |
| 24 | **task** [tɑːsk] | Aufgabe, Arbeit |
| | **column** [ˈkɒləm] | Spalte, Kolumne |
| | **tension** [ˈtenʃn] | Spannung |
| | **blow** [bləʊ] | blasen |
| | **cloth** [klɒθ] | Stoff, Tuch, Lappen |
| | **waste** [weɪst] | Abfall |
| | **leakage** [ˈliːkɪdʒ] | Leck(stelle) |
| | **compressed** [kəmˈprest] | Druck- |
| | **interval** [ˈɪntəvl] | Abstand |
| | **surplus** [ˈsɜːpləs] | überschüssig |
| | **timing belt** [ˈtaɪmɪŋ belt] | Antriebsriemen, Synchronriemen |
| | **shift** [ʃɪft] | Schicht |
| | **ventilator** [ˈventɪleɪtə] | Ventilator |
| | **blow out** [ˌbləʊ ˈaʊt] | ausblasen |
| | **filter mat** [ˈfɪltə mæt] | Filter(einsatz) |
| | **electronic card** [ˌɪlekˈtrɒnɪk kɑːd] | elektronische Karte |
| | **water pipe** [ˈwɔːtə paɪp] | Wasserrohr |
| | **hose** [həʊz] | Schlauch |
| | **check for** [ˈtʃek fə] | nachprüfen |
| 25 | **technical** [ˈteknɪkl] | technisch, Fach- |
| | **occupation** [ˌɒkjuˈpeɪʃn] | Beruf, Tätigkeit |
| | **well established** [ˌwel ɪˈstæblɪʃt] | etabliert, gut eingeführt |
| | **take up** [ˌteɪk ˈʌp] | anfangen mit |
| | **train for** [ˈtreɪn fə] | sich ausbilden lassen für |
| | **increasingly** [ɪnˈkriːsɪŋli] | zunehmend |
| | **recognize** [ˈrekəgnaɪz] | (an)erkennen |
| | **value** [ˈvæljuː] | Wert, Nutzen |
| | **employ** [ɪmˈplɔɪ] | ein-, anstellen, beschäftigen |
| | **dedicated** [ˈdedɪkeɪtɪd] | engagiert |
| | **reflect** [rɪˈflekt] | (sich) widerspiegeln |
| | **electrician** [ɪˌlekˈtrɪʃn] | Elektriker/in, Elektrotechniker/in |

| | | |
|---|---|---|
| | **mechanic** [mɪ'kænɪk] | Mechaniker/in |
| | **lecturer** ['lektʃərə] | Dozent/in |
| | **scheme** [ski:m] | Plan, Programm |
| | **be similar to** [bi 'sɪmələ tə] | ähneln |
| | **unfortunately** [ʌn'fɔ:tʃənətli] | unglücklicherweise, leider |
| | **route** [ru:t] | Route, Weg |
| | **HNC (Higher National Certificate)** [ˌhaɪə ˌnæʃnəl sə'tɪfɪkət] | staatliches Abschlusszeugnis |
| | **study for** ['stʌdi fə] | studieren |
| | **college of further education** [ˌkɒlɪdʒ əv ˌfɜ:ðər ˌedʒu'keɪʃn] | Berufsschule, Fachhochschule (für die Berufsausbildung) |
| | **roughly** ['rʌfli] | ungefähr, etwa |
| | **be equivalent to** [ɪ'kwɪvələnt tə] | entsprechen |
| | **vocational college** [vəʊˌkeɪʃənl 'kɒlɪdʒ] | Berufs(ober)schule |
| | **last for** ['lɑ:st fə] | dauern |
| | **depending on** [dɪ'pendɪŋ ɒn] | je nach(dem) |
| | **full/part-time** [ˌfʊl, ˌpɑ:t 'taɪm] | ganz-/halbtags |
| | **course** [kɔ:s] | Kurs, Lehrgang, Unterricht |
| | **employer** [ɪm'plɔɪə] | Arbeitgeber/in |
| | **strength** [streŋθ] | Stärke |
| | **depth** [depθ] | Tiefe |
| | **breadth** [bredθ] | Breite |
| | **skill** [skɪl] | Fertigkeit, Fähigkeit |
| | **such as** ['sʌtʃ əz] | wie (zum Beispiel) |
| | **pneumatics** [nju:'mætɪks] | Pneumatik |
| | **hydraulics** [haɪ'drɔ:lɪks] | Hydraulik |
| | **PLC (Programmable Logic Controller)** [ˌprəʊgræməbl ˌlɒdʒɪk kən'trəʊlə] | speicherprogrammierbare Steuerung |
| | **computer-aided** [kəm'pju:tər eɪdɪd] | computergesteuert, -gestützt |
| | **sensor** ['sensə] | Sensor, Messgerät, Melder, (Mess-)Fühler |
| | **systems monitoring** [ˌsɪstəmz 'mɒnɪtərɪŋ] | Produktionskontrolle |
| | **in addition to this** [ɪn ə'dɪʃn tə ðɪs] | darüber hinaus |
| | **communication** [kəˌmju:nɪ'keɪʃn] | Kommunikation |
| | **project management** [ˌprɒdʒekt 'mænɪdʒmənt] | Projektmanagement |
| | **manufacture** [ˌmænju'fæktʃə] | herstellen, produzieren, verarbeiten |
| | **swap** [swɒp] | (ver)tauschen |

## Unit 3

| | | |
|---|---|---|
| 26 | **capacitance meter** [kə'pæsɪtəns mi:tə] | Kapazitätsmesser |
| | **function generator** ['fʌŋkʃn dʒenəreɪtə] | Funktionsgenerator |
| | **wattmeter** ['wɒtmi:tə] | Wattmeter |
| | **test light** ['test laɪt] | Prüflampe |
| | **determine** [dɪ'tɜ:mɪn] | feststellen |
| | **presence** ['prezns] | Anwesenheit |
| | **absence** ['æbsəns] | Abwesenheit |
| | **electric** [ɪ'lektrɪk] | elektrisch |
| | **current** ['kʌrənt] | Strom |
| | **resistance** [rɪ'zɪstəns] | Widerstand |
| | **generate** ['dʒenəreɪt] | erzeugen |
| | **waveform** ['weɪvfɔ:m] | Wellenform, Kurve |
| | **newly-designed** [ˌnju:li dɪ'zaɪnd] | neu konstruiert |
| | **circuitry** ['sɜ:kɪtri] | Schaltung, Kreislauf, (Strom-)Kreis |
| | **find out** [ˌfaɪnd 'aʊt] | herausfinden, feststellen |
| | **electronic** [ˌɪlek'trɒnɪk] | elektronisch |
| | **faulty** ['fɔ:lti] | fehlerhaft, defekt |
| | **rate** [reɪt] | Rate |
| | **energy** ['enədʒi] | Energie |
| | **given** ['gɪvn] | gegeben |
| | **circuit** ['sɜ:kɪt] | Schaltung, (Strom-)Kreis |
| | **capacitor** [kə'pæsɪtə] | Kondensator |
| | **advice** [əd'vaɪs] | Rat, Ratschlag |
| | **put sb through** [ˌpʊt 'θru:] | jdn durchstellen, jdn verbinden |
| 27 | **have a minute** [həv ə 'mɪnɪt] | eine Minute (Zeit) haben |
| | **specs (specifications)** [speks, ˌspesɪfɪ'keɪʃnz] | genaue Angaben, Spezifikationen |
| | **handy** ['hændi] | zur Hand, parat |
| | **hang on** [ˌhæŋ 'ɒn] | dranbleiben |
| | **screen** [skri:n] | Bildschirm |
| | **digital multimeter** [ˌdɪdʒɪtl ˌmʌlti'mi:tə] | digitales Vielfachmessinstrument |
| | **capacitance** [kə'pæsɪtəns] | Kapazität |
| | **range** [reɪndʒ] | Bandbreite, Bereich |
| | **selection** [sɪ'lekʃn] | Auswahl |
| | **broad** [brɔ:d] | breit |
| | **narrow** ['nærəʊ] | eng |
| | **audible continuity** ['ɔ:dəbl ˌkɒntɪ'nju:əti] | hörbarer Durchgang |
| | **transistor** [træn'zɪstə] | Transistor |
| | **diode** ['daɪəʊd] | Diode |
| | **test facility** [test fə'sɪləti] | Prüf-, Testvorrichtung |
| | **overload protection** [ˌəʊvələʊd prə'tekʃn] | Überlastungsschutz |
| | **auto-ranging** ['ɔ:təʊ reɪndʒɪŋ] | Selbststerstreckung |
| | **data hold function** [ˌdeɪtə həʊld 'fʌŋkʃn] | Funktion zur Datenspeicherung |
| | **conform to** [kən'fɔ:m tə] | entsprechen |
| | **overvoltage** [ˌəʊvə'vəʊltɪdʒ] | Überlastung |
| | **come with** ['kʌm wɪð] | geliefert werden mit |
| | **(instruction) manual** [ɪnˌstrʌkʃn 'mænjuəl] | Handbuch |
| | **probe** [prəʊb] | Sonde |
| | **spare** [speə] | Ersatz- |
| | **battery** ['bætəri] | Batterie |
| | **holster** ['həʊlstə] | Halfter, Tragetasche |
| | **battery life** ['bætəri laɪf] | Batterielebensdauer |
| | **sometime** ['sʌmtaɪm] | irgendwann |

| | | |
|---|---|---|
| | **worry about** ['wʌri əbaʊt] | sich Gedanken machen wegen |
| | **weight** [weɪt] | Gewicht |
| | **go for sth** ['gəʊ fə] | etw nehmen, sich für etw entscheiden |
| 28 | **in pairs** [ɪn 'peəz] | paarweise |
| | **role-play** ['rəʊl pleɪ] | als Rollenspiel durchführen |
| | **each other** [i:tʃ 'ʌðə] | einander, gegenseitig |
| | **sheet** [ʃi:t] | Blatt |
| | **contact** ['kɒntækt] | Kontakt |
| | **connect (to)** [kə'nekt tə] | verbinden (mit) |
| | **hold the line** [həʊld ðə 'laɪn] | dranbleiben |
| | **apologize** [ə'pɒlədʒaɪz] | sich entschuldigen |
| | **available** [ə'veɪləbl] | (am Telefon) zu sprechen |
| | **take a message** [ˌteɪk ə 'mesɪdʒ] | etw ausrichten, eine Nachricht entgegen-nehmen |
| | **promise** ['prɒmɪs] | versprechen, zusagen |
| | **get back to sb about sth** [ˌget 'bæk tə əbaʊt] | sich wieder bei jdm wegen etw melden |
| | **let sb know** [ˌlet 'nəʊ] | jdm Bescheid geben/sagen |
| | **difficulty** ['dɪfɪkəlti] | Schwierigkeit |
| | **read back** [ˌri:d 'bæk] | (noch einmal) vorlesen |
| | **hyphen** ['haɪfn] | Bindestrich |
| | **underscore** [ˌʌndə'skɔ:] | unterstreichen |
| | **dot** [dɒt] | Punkt |
| | **digit** ['dɪdʒɪt] | (einstellige) Zahl, Ziffer |
| | **separately** ['seprətli] | getrennt |
| | **ext. (extension)** [ɪk'stenʃn] | Durchwahl, Apparat |
| 28p | **approx. (approximate)** [ə'prɒksɪmət] | ungefähr, etwa, circa |
| | **input impedance** [ˌɪnpʊt ɪm'pi:dns] | Eingangsimpedanz |
| | **Z-axis** ['zed æksɪs] | Z-Achse |
| | **input** ['ɪnpʊt] | Eingang, Aufnahme |
| | **sensitivity** [ˌsensə'tɪvəti] | Empfindlichkeit |
| | **vertical(ly)** ['vɜ:tɪkl, -i] | senkrecht, vertikal |
| | **feature** ['fi:tʃə] | Merkmal |
| | **Dual Channel High sensitivity** [ˌdju:əl ˌtʃænl haɪ ˌsensə'tɪvəti] | Hohe Zweikanalempfind-lichkeit |
| | **synchronization** [ˌsɪŋkrənaɪ'zeɪʃn] | Synchronisierung |
| | **ALT Triggering Function** [ˌeɪ el 'ti: 'trɪgərɪŋ fʌŋkʃn] | ALT-Auslösefunktion |
| | **output** ['aʊtpʊt] | Ausgang, -gabe |
| | **CRT** [ˌsi: ɑ: 'ti:] | CRT-Display, Flachbildschirm |
| | **type** [taɪp] | Art, Typ |
| | **inch** [ɪntʃ] | Inch, Zoll |
| | **rectangular** [rek'tæŋgjələ] | rechteckig |
| | **internal graticule** [ɪnˌtɜ:nl 'grætɪkju:l] | internes Netz, Gitter, Koordinatensystem |
| | **bandwidth** ['bændwɪdθ] | Bandbreite |
| | **rise time** ['raɪz taɪm] | Anstiegszeit |
| | **coupling** ['kʌplɪŋ] | Kopplung |
| | **mode** [məʊd] | Art, Modus |
| | **horizontal(ly)** [ˌhɒrɪ'zɒntl, -i] | horizontal, waagerecht |
| | **sweep time** ['swi:p taɪm] | Abtastzeit |
| | **uncalibrated** [ˌʌn'kælɪbreɪtɪd] | nicht kalibriert |
| | **trigger** ['trɪgə] | Trigger, Auslöser |
| | **trigger mode** ['trɪgə məʊd] | Trigger-Art |
| | **trigger source** ['trɪgə sɔ:s] | Trigger-Quelle |
| | **trigger slope** ['trɪgə sləʊp] | Trigger-Empfindlichkeit |
| | **x-y operation** [ˌeks 'waɪ ˌɒpə'reɪʃn] | X-Y-Tätigkeit |
| | **x-axis** ['eks æksɪs] | X-Achse |
| | **phase error** ['feɪz erə] | Phasenverschiebung |
| | **trigger signal output** [ˌtrɪgə sɪgnəl 'aʊtpʊt] | Trigger-Signalausgang |
| | **calibrator** ['kælɪbreɪtə] | Kalibrierer, Kalibrator |
| | **squarewave** ['skweəweɪv] | Rechteckwelle |
| | **power source** ['paʊə sɔ:s] | Netzanschluss |
| | **accessory** [ək'sesəri] | Zubehör |
| | **power cord** ['paʊə kɔ:d] | Netz-, Stromkabel |
| | **dimension** [dɪ'menʃn] | Abmessung, Maß |
| 29 | **make notes** [ˌmeɪk 'nəʊts] | (sich) Notizen machen |
| | **experienced** [ɪk'spɪəriənst] | erfahren |
| 30 | **up to standard** [ʌp tə 'stændəd] | der Norm entsprechend |
| | **replace** [rɪ'pleɪs] | ersetzen |
| | **suitable** ['su:təbl] | geeignet, passend |
| | **accurate** ['ækjərət] | genau, richtig |
| | **switch on** [ˌswɪtʃ 'ɒn] | an-, einschalten |
| | **overnight** [ˌəʊvə'naɪt] | über Nacht |
| | **try out** [ˌtraɪ 'aʊt] | (aus)probieren, testen |
| | **model** ['mɒdl] | Modell, Gerät |
| | **exhibition** [ˌeksɪ'bɪʃn] | Ausstellung |
| | **tidy up** [ˌtaɪdi 'ʌp] | aufräumen, putzen |
| 31 | **mathematical** [ˌmæθə'mætɪkl] | mathematisch |
| | **operation** [ˌɒpə'reɪʃn] | Operation |
| | **greater/larger than** ['greɪtə, 'lɑ:dʒə ðən] | größer als |
| | **smaller than** ['smɔ:lə ðən] | kleiner als |
| | **equals** ['i:kwəlz] | ist gleich |
| | **plus** [plʌs] | plus |
| | **minus** ['maɪnəs] | minus |
| | **times** [taɪmz] | mal |
| | **multiplied by** ['mʌltɪplaɪd baɪ] | multipliziert mit |
| | **divided by** [dɪ'vaɪdɪd baɪ] | geteilt durch |
| | **note** [nəʊt] | beachten |
| | **pronounce** [prə'naʊns] | aussprechen |
| | **arithmetic** [ə'rɪθmətɪk] | Rechnen, Arithmetik |
| | **result in** [rɪ'zʌlt ɪn] | ergeben |
| | **calculation** [ˌkælkju'leɪʃn] | Rechenart, -operation |
| | **write out** [ˌraɪt 'aʊt] | ausschreiben |
| | **squared** [skweəd] | hoch zwei |
| | **cubed** [kju:bd] | hoch drei |
| | **square root of** ['skweə ru:t əv] | Quadratwurzel |
| | **to the power of** [tə ðə 'paʊər əv] | hoch |

| | | |
|---|---|---|
| 31p | **read out** [ˌriːd 'aʊt] | (laut) vorlesen |
| | **equation** [ɪ'kweɪʒn] | Gleichung |
| 32 | **definition** [ˌdefɪ'nɪʃn] | Definition |
| | **amp (ampere)** [æmp, 'æmpeə] | Ampere |
| | **volt** [vəʊlt] | Volt |
| | **Ohm's law** [əʊmz 'lɔː] | das Ohmsche Gesetz |
| | **define** [dɪ'faɪn] | bestimmen, festlegen |
| | **maintain** [meɪn'teɪn] | einhalten |
| | **conductor** [kən'dʌktə] | Leiter |
| | **surface** ['sɜːfɪs] | (Ober-)Fläche |
| | **potential** [pə'tenʃl] | Potenzial |
| | **control** [kən'trəʊl] | kontrollieren, regeln |
| | **amount** [ə'maʊnt] | Menge |
| | **watt** [wɒt] | Watt |
| | **wattage** ['wɒtɪdʒ] | Wattleistung |
| | **adapted from** [ə'dæptɪd frəm] | adaptiert von |
| | **pie chart** ['paɪ tʃɑːt] | Kreisdiagramm |
| | **remaining** [rɪ'meɪnɪŋ] | restlich, verbleibend |
| | **express** [ɪk'spres] | ausdrücken |
| | **formula (pl. formulae)** ['fɔːmjələ, -liː] | Formel |
| | **segment** ['segmənt] | Teilstück, Segment |
| 33 | **graph** [grɑːf] | Kurve, grafische Darstellung, Grafik |
| | **drop** [drɒp] | fallen, sinken |
| | **flatten out** [ˌflætn 'aʊt] | (sich) abflachen |
| | **fluctuate** ['flʌktʃueɪt] | schwanken |
| | **hit** [hɪt] | erreichen |
| | **level off** [ˌlevl 'ɒf] | sich einpendeln |
| | **peak** [piːk] | gipfeln, den Höchstwert erreichen |
| | **plunge** [plʌndʒ] | (ab)stürzen, absacken |
| | **remain** [rɪ'meɪn] | bleiben |
| | **constant** ['kɒnstənt] | konstant |
| | **shoot up** [ˌʃuːt 'ʌp] | in die Höhe schießen |
| | **drastic(ally)** ['dræstɪkli] | drastisch |
| | **gradual(ly)** ['grædʒuəl, -li] | allmählich |
| | **sharp(ly)** ['ʃɑːp, -li] | rasch, schnell, stark |
| | **steady (-ily)** ['stedɪli, -li] | gleich-, regelmäßig, stetig, ständig |
| | **suddenly** ['sʌdnli] | plötzlich |
| 34 | **characteristic curve** [ˌkærəktə'rɪstɪk kɜːv] | Kennlinie |
| | **translation** [træns'leɪʃn] | Übersetzung |
| | **reversed-biased** [rɪ'vɜːst baɪəst] | in Sperrrichtung |
| | **reverse voltage** [rɪˌvɜːs 'vəʊltɪdʒ] | Sperrspannung |
| | **forward-biased** ['fɔːwəd baɪəst] | im Durchlassbereich |
| | **forward voltage** [ˌfɔːwəd 'vəʊltɪdʒ] | Durchlassspannung |
| | **fault** [fɔːlt] | Fehler, Defekt |
| 34t | **diagnose** ['daɪəgnəʊz] | diagnostizieren |
| | **break down** [ˌbreɪk 'daʊn] | durchbrechen |
| 34 | **curve** [kɜːv] | Kurve |
| | **swing** [swɪŋ] | schwingen |
| | **curve back/round** ['kɜːv bæk, raʊnd] | sich zurückbiegen, krümmen |
| | **double back on itself** [ˌdʌbl 'bæk ɒn] | kehrtmachen, zurückkehren |
| | **mirror** ['mɪrə] | (wider)spiegeln |
| | **mirror image** [ˌmɪrə 'ɪmɪdʒ] | Spiegelbild |
| | **zero/neutral point** ['zɪərəʊ, 'njuːtrəl pɔɪnt] | Nullpunkt |
| 35 | **lead the way** [liːd ðə 'weɪ] | anführen, vorangehen |
| | **up until** [ʌp ən'tɪl] | bis (zu) |
| | **car owner** ['kɑːr əʊnə] | Autobesitzer/in |
| | **be keen to** [bi 'kiːn tə] | erpicht sein auf, sehr interessiert sein an |
| | **vehicle** ['viːəkl] | Fahrzeug |
| | **proud** [praʊd] | stolz |
| | **guarantee** [ˌgærən'tiː] | Garantie |
| | **look around** [ˌlʊk ə'raʊnd] | (sich) umsehen |
| | **stand one's ground against** [ˌstænd wʌnz 'graʊnd əgənst] | seine Position behaupten gegen(über) |
| | **automobile manufacturing industry** [ˌɔːtəməbiːl ˌmænju'fæktʃərɪŋ ɪndəstri] | Autoindustrie |
| | **mind** [maɪnd] | hier: Meinung |
| | **improvement** [ɪm'pruːvmənt] | Verbesserung |
| | **along with** [ə'lɒŋ wɪð] | zusammen mit |
| | **production smoothing** [prə'dʌkʃn smuːðɪŋ] | Produktionsglättung |
| | **eliminate** [ɪ'lɪmɪneɪt] | eliminieren, beseitigen |
| | **automobile** ['ɔːtəməbiːl] | Auto(mobil) |
| | **work out** [ˌwɜːk 'aʊt] | ausarbeiten, sich ausdenken |
| | **philosophy** [fɪ'lɒsəfi] | Philosophie |
| | **be aimed at** ['eɪmd ət] | (ab)zielen auf, anstreben |
| | **top-quality** [ˌtɒp 'kwɒləti] | Spitzenqualitäts- |
| | **reasonable** ['riːznəbl] | angemessen, günstig |
| | **eventually** [ɪ'ventʃuəli] | schließlich, zum Schluss |
| | **lean manufacturing** [liːn ˌmænju'fæktʃərɪŋ] | schlanke Produktion |
| | **human** ['hjuːmən] | menschlich |
| | **intelligence** [ɪn'telɪdʒəns] | Intelligenz |
| | **programme into** ['prəʊgræm ɪntə] | einprogrammieren |
| | **automated** ['ɔːtəmeɪtɪd] | automatisiert |
| | **unload** [ˌʌn'ləʊd] | entladen |
| | **automatic(ally)** [ˌɔːtə'mætɪk, -kli] | automatisch |
| | **detect** [dɪ'tekt] | entdecken |
| | **defective** [dɪ'fektɪv] | fehlerhaft |
| | **signal for** ['sɪgnəl fə] | herbeirufen |
| | **index card** ['ɪndeks kɑːd] | Karteikarte |
| | **record card** ['rekɔːd kɑːd] | Karteikarte |
| | **ensure** [ɪn'ʃʊə] | sicherstellen, gewährleisten |

| | | |
|---|---|---|
| | **in response to** [ɪn rɪ'spɒns tə] | als Reaktion auf, entsprechend |
| | **demand** [dɪ'mɑ:nd] | Nachfrage, Bedarf |
| | **workstation** ['wɜ:ksteɪʃn] | Arbeitsplatz (mit Computer) |
| | **avoid** [ə'vɔɪd] | (ver)meiden |
| | **inadvertent** [ˌɪnəd'vɜ:tənt] | unbeabsichtigt, versehentlich |
| | **error** ['erə] | Fehler, Irrtum, Versehen |
| | **defect** [dɪ'fekt] | Defekt |
| | **malfunction** [ˌmæl'fʌŋkʃn] | Fehlfunktion |
| | **occur** [ə'kɜ:] | vorkommen, passieren, eintreten |
| | **choice** [tʃɔɪs] | Wahl, Auswahl |
| | **incorrect** [ˌɪnkə'rekt] | falsch, unrichtig |
| | **prevent** [prɪ'vent] | (ver)hindern, verhüten |
| | **damage** ['dæmɪdʒ] | Schaden, Beschädigung |
| | **seeming(ly)** ['si:mɪŋ, -li] | scheinbar, anscheinend |
| | **improve** [ɪm'pru:v] | verbessern |
| | **on a daily basis** [ɒn ə ˌdeɪli 'beɪsɪs] | täglich |
| | **long term** [ˌlɒŋ 'tɜ:m] | langfristig, Langzeit- |
| | **adapt** [ə'dæpt] | anpassen, adaptieren |
| | **fit** [fɪt] | passen (zu), entsprechen |
| | **environment** [ɪn'vaɪrənmənt] | Umwelt, Umfeld |
| | **suggest** [sə'dʒest] | vorschlagen |
| | **achieve** [ə'tʃi:v] | erreichen, erlangen, erzielen |
| | **balance** ['bæləns] | Gleichgewicht, Ausgleich |
| | **intelligent** [ɪn'telɪdʒənt] | intelligent |
| | **discussion** [dɪ'skʌʃn] | Diskussion, Gespräch |

## Unit 4

| | | |
|---|---|---|
| 36 | **diac** ['daiək] | Diac |
| | **switch** [swɪtʃ] | Schalter |
| | **thyristor** [θaɪ'rɪstə] | Thyristor |
| | **give off** [ˌgɪv 'ɒf] | ausstrahlen |
| | **act like** ['ækt laɪk] | arbeiten wie |
| | **be made up of** [ˌmeɪd 'ʌp əv] | zusammengesetzt sein aus, bestehen aus |
| | **layer** ['leɪə] | Schicht |
| | **alternating** ['ɔ:ltɜ:neɪtɪŋ] | abwechselnd |
| | **mainly** ['meɪnli] | hauptsächlich, vor allem |
| | **involve** [ɪn'vɒlv] | beteiligen |
| | **limit** ['lɪmɪt] | begrenzen, beschränken |
| | **made from** ['meɪd frəm] | hergestellt aus |
| | **carbon** ['kɑ:bən] | Kohlenstoff |
| | **semi-conducting** [ˌsemi kən'dʌktɪŋ] | Halbleiter- |
| | **silicon** ['sɪlɪkən] | Silizium |
| | **amplifier** ['æmplɪfaɪə] | Verstärker |
| | **turn on/off** [ˌtɜ:n 'ɒn, 'ɒf] | ein-/ausschalten |
| | **bi-directional** [ˌbaɪ də'rekʃənl] | Zweirichtungs- |
| | **triac** ['traɪæk] | Triac |
| | **dissipation** [ˌdɪsɪ'peɪʃn] | Ableitung |
| | **pin** [pɪn] | heften |
| | **in series** [ɪn 'sɪəri:z] | in Reihe geschaltet |
| | **in parallel** [ɪn 'pærəlel] | parallel geschaltet |
| | **solve** [sɒlv] | lösen |
| | **fix** [fɪks] | reparieren, in Ordnung bringen |
| | **scan** [skæn] | überfliegen, durchsehen |
| | **ammeter** ['æmɪtə] | Amperemeter |
| | **mains tester** ['meɪnz teste] | Strommesser |
| | **ohmmeter** ['əʊmmi:tə] | Ohmmeter |
| | **trimpot** ['trɪmpɒt] | Trimmpotentiometer |
| 37 | **factory floor** ['fæktəri flɔ:] | Produktion(sfläche) |
| | **bother sb** ['bɒðə] | jdn stören |
| | **trouble** ['trʌbl] | Schwierigkeiten, Mühe |
| | **packaging robot** ['pækɪdʒɪŋ rəʊbɒt] | Verpackungsroboter |
| | **repair brief** [rɪ'peə bri:f] | Reparaturanleitung, Serviceunterlagen |
| | **specifications** [ˌspesɪfɪ'keɪʃnz] | genaue Angaben, Spezifikationen |
| | **mechanics** [mɪ'kænɪks] | Mechanik, Ablauf |
| | **stuck** [stʌk] | fest-, eingeklemmt |
| | **not in any way** [nɒt ɪn ˌeni 'weɪ] | in keiner Weise |
| | **electrics** [ɪ'lektrɪks] | Elektrik |
| | **short-circuit** [ˌʃɔ:t 'sɜ:kɪt] | einen Kurzschluss auslösen |
| | **equally** ['i:kwəli] | gleichermaßen |
| | **fault-finding** ['fɔ:lt faɪndɪŋ] | Fehlersuche |
| | **some months back** [sʌm ˌmʌnθs 'bæk] | vor einigen Monaten |
| | **come loose** [kʌm 'lu:s] | sich lockern |
| | **circuit board** ['sɜ:kɪt bɔ:d] | Platine, Schalttafel |
| | **send out** [ˌsend 'aʊt] | verschicken, aussenden |
| | **bet** [bet] | wetten |
| | **get back to sb** [ˌget 'bæk tə] | jdn zurückrufen |
| | **hang off** [ˌhæŋ 'ɒf] | (lose) hängen an |
| | **come up to** [ˌkʌm 'ʌp tə] | herkommen |
| | **summarize** ['sʌməraɪz] | zusammenfassen |
| | **fill in** [ˌfɪl 'ɪn] | ausfüllen |
| | **time sheet** ['taɪm ʃi:t] | Stunden-, Arbeitszettel |
| | **moving part** [ˌmu:vɪŋ 'pɑ:t] | bewegliches Teil |
| | **session** ['seʃn] | Sitzung |
| | **re-solder** [ˌri: 'səʊldə] | anlöten |
| 38 | **ending** ['endɪŋ] | Endung, Ende (der Geschichte) |
| | **electronics** [ˌɪlek'trɒnɪks] | Elektronik |
| | **mechanical** [mɪ'kænɪkl] | mechanisch |
| | **missing** ['mɪsɪŋ] | fehlend, nicht vorhanden |
| | **corroded** [kə'rəʊdɪd] | korrodiert, verrostet |
| | **make sense** [meɪk 'sens] | sinnvoll sein |
| | **as far as ... is concerned** [əz ˌfɑ:r əz ɪz kən'sɜ:nd] | was ... anbetrifft |
| | **scrape** [skreɪp] | kratzen, schaben, reiben |
| 39 | **proper** ['prɒpə] | richtig, ordnungsgemäß |
| | **get on with sb** [ˌget 'ɒn wɪð] | sich mit jdm verstehen |
| | **respect** [rɪ'spekt] | respektieren |
| | **condition** [kən'dɪʃn] | Bedingung |

| | | |
|---|---|---|
| | **assembly** [ə'sembli] | Montage |
| | **motion** ['məʊʃn] | Bewegung |
| | **programmer** ['prəʊgræmə] | Programmierer/in |
| | **robotic** [rəʊ'bɒtɪk] | Roboter- |
| | **repeat** [rɪ'pi:t] | wiederholen |
| | **movement** ['mu:vmənt] | Bewegung |
| | **over and over again** [ˌəʊvər ənd ˌəʊvər ə'gen] | immer wieder |
| | **guide** [gaɪd] | führen, lenken |
| | **handheld** ['hændheld] | Hand-(Gerät), tragbar |
| | **controller** [kən'trəʊlə] | Steuer-, Bedienungsgerät |
| | **memory** ['meməri] | Speicher |
| | **assembly line** [ə'sembli laɪn] | Montageband |
| 40 | **impression** [ɪm'preʃn] | Eindruck |
| | **particularly** [pə'tɪkjələli] | besonders |
| | **interrupt** [ˌɪntə'rʌpt] | unterbrechen |
| | **request** [rɪ'kwest] | Bitte, Anfrage, Wunsch, Anweisung |
| | **based on** ['beɪst ɒn] | auf der Grundlage von |
| | **gears** [gɪəz] | Getriebe |
| 41 | **expect** [ɪk'spekt] | erwarten, schätzen, annehmen |
| | **give a talk** [gɪv ə 'tɔ:k] | einen Vortrag halten |
| 41t | **representative** [ˌreprɪ'zentətɪv] | Vertreter/in, Repräsentant/in |
| | **distribute** [dɪ'strɪbju:t] | vertreiben, (aus)liefern |
| | **all over the world** [ɔ:l ˌəʊvə ðə 'wɜ:ld] | in der ganzen Welt |
| | **get a picture of** [get ə 'pɪktʃər əv] | sich ein Bild machen |
| | **overview** ['əʊvəvju:] | Überblick, -sicht |
| | **one by one** [ˌwʌn baɪ 'wʌn] | einer nach dem anderen |
| | **not mind** [nɒt 'maɪnd] | nichts dagegen haben |
| | **interruption** [ˌɪntə'rʌpʃn] | Unterbrechung |
| 41 | **semiconductor** [ˌsemikən'dʌktə] | Halbleiter |
| | **doping** ['dəʊpɪŋ] | Doping, Verunreinigen |
| | **electron** [ɪ'lektrɒn] | Elektron |
| | **germanium** [dʒɜ:'meɪniəm] | Germanium |
| | **outer** ['aʊtə] | äußere, Außen- |
| | **orbital** ['ɔ:bɪtl] | Umlaufbahn |
| | **impurity** [ɪm'pjʊərəti] | Verunreinigung, Unreinheit |
| | **crystal** ['krɪstl] | Kristall |
| | **covalent bond** [ˌkəʊ'veɪlənt bɒnd] | Bindung durch gemeinsame Elektronen |
| | **neighbouring** ['neɪbərɪŋ] | angrenzend, benachbart |
| | **atom** ['ætəm] | Atom |
| | **create** [kri'eɪt] | bilden, erzeugen |
| | **lattice** ['lætɪs] | Gitter |
| | **behaviour** [bɪ'heɪvjə] | Verhalten |
| | **turn sth into** ['tɜ:n ɪntə] | etw machen zu |
| | **at the heart of** [ət ðə 'hɑ:t əv] | im Zentrum von |
| | **device** [dɪ'vaɪs] | Vorrichtung, Gerät |
| | **quartz** [kwɔ:ts] | Quarz |
| 42 | **computerized** [kəm'pju:təraɪzd] | Computer-, computerisiert |
| | **wave** [weɪv] | Welle |
| | **microprocessor** ['maɪkrəʊprəʊsesə] | Mikroprozessor |
| | **catalogue** ['kætəlɒg] | Katalog |
| | **electrolytic capacitor** [ɪlektrəˌlɪtɪk kə'pæsɪtə] | Elektrolytkondensator |
| | **transformer** [træns'fɔ:mə] | Transformator |
| | **secondary** ['sekəndri] | sekundär, Neben- |
| | **bridge rectifier** [brɪdʒ 'rektɪfaɪə] | Brückengleichrichter |
| 43 | **circuit diagram** ['sɜ:kɪt daɪəgræm] | Schaltbild |
| | **neat(ly)** ['ni:t, -li] | ordentlich, sauber |
| | **junction** ['dʒʌŋkʃn] | Kreuzung, Verbindungsstelle, Knoten |
| | **omit** [ə'mɪt] | aus-, weglassen |
| | **power supply** ['paʊə səplaɪ] | Netz-, Stromanschluss, Spannungsversorgung |
| | **include** [ɪn'klu:d] | angeben |
| | **label** ['leɪbl] | beschriften, kennzeichnen |
| | **straight** [streɪt] | gerade |
| | **blob** [blɒb] | Punkt |
| | **complex** ['kɒmpleks] | kompliziert, komplex |
| | **arrange** [ə'reɪndʒ] | arrangieren, anordnen |
| 44 | **conduct** [kən'dʌkt] | leiten |
| | **insulate** ['ɪnsjuleɪt] | isolieren, abschirmen |
| | **attach to** [ə'tætʃt tə] | befestigen an, verbinden mit |
| | **fixed to** ['fɪkst tə] | befestigt an |
| | **joined to** ['dʒɔɪnd tə] | verbunden mit |
| | **left-hand side** [ˌleft hænd 'saɪd] | linke Seite |
| | **go off to** [ˌgəʊ 'ɒf tə] | abgehen zu |
| 45 | **pick and place robot** [ˌpɪk ən 'pleɪs rəʊbɒt] | Bestückungs-/ Montageroboter |
| | **cell** [sel] | Zelle |
| | **servant** ['sɜ:vənt] | Diener/in |
| | **entertain** [ˌentə'teɪn] | unterhalten |
| | **everyday life** [ˌevrideɪ 'laɪf] | Alltag(sleben) |
| | **feel comfortable with** [fi:l 'kʌmftəbl wɪð] | sich gut fühlen mit |
| | **roboticist** [rəʊ'bɒtɪsɪst] | Roboteringenieur/in |
| | **acceptable** [ək'septəbl] | annehmbar, akzeptabel |
| | **in order to** [ɪn 'ɔ:də tə] | um zu |
| | **set up** [ˌset 'ʌp] | veranstalten |
| | **volunteer** [ˌvɒlən'tɪə] | Freiwillige/r |
| | **resident** ['rezɪdənt] | wohnhaft, ansässig |
| | **light up** [ˌlaɪt 'ʌp] | aufleuchten |
| | **relate to** [rɪ'leɪt tə] | eine Beziehung haben zu, reagieren auf |
| | **pure(ly)** ['pjʊə, -li] | rein |
| | **conference** ['kɒnfərəns] | Konferenz, Tagung |
| | **challenge** ['tʃælɪndʒ] | An-, Herausforderung, Aufgabe |

| | |
|---|---|
| **come up with** [ˌkʌm 'ʌp wɪð] | erfinden, sich ausdenken |
| **actually** ['æktʃuəli] | wirklich, tatsächlich, eigentlich |
| **communicate** [kə'mju:nɪkeɪt] | sich verständigen, kommunizieren |
| **enormous** [ɪ'nɔ:məs] | enorm, gewaltig |
| **speech** [spi:tʃ] | Sprache |
| **gesture** ['dʒestʃə] | Geste(n), Gestik |
| **make eye contact** [meɪk 'aɪ kɒntækt] | Blickkontakt herstellen |
| **have an awareness of sth** [həv ən ə'weənəs əv] | sich einer Sache bewusst sein |
| **personal space** [ˌpɜ:sənl 'speɪs] | Diskretionsabstand |
| **sort** [sɔ:t] | Art, Sorte |
| **thought** [θɔ:t] | Gedanke |
| **paragraph** ['pærəgrɑ:f] | Absatz, Abschnitt |

## Unit 5

| | | |
|---|---|---|
| 46 | **installation** [ˌɪnstə'leɪʃn] | Installation, Auf-, Einbau, Montage |
| | **X-ray** ['eks reɪ] | Röntgenstrahl; röntgen |
| | **tube** [tju:b] | Tube |
| | **toothpaste** ['tu:θpeɪst] | Zahnpasta |
| | **fibre optic cable** [ˌfaɪbər ɒptɪk 'keɪbl] | Glasfaserkabel |
| | **compressed-air pistol** [kəmˌprest eə 'pɪstl] | Druckluftpistole |
| | **compression moulding** [kəm'preʃn məʊldɪŋ] | Druckspritz- |
| | **compression moulding unit** [kəm'preʃn məʊldɪŋ ju:nɪt] | Druckspritzmaschine |
| | **elevator** ['elɪveɪtə] | Aufzug |
| | **mandrel** ['mændrəl] | Dorn, Spindel |
| | **operating panel** ['ɒpəreɪtɪŋ pænl] | Betriebs-, Schalttafel |
| | **tube body maker** [tju:b 'bɒdi meɪkə] | Tubenkörpermaschine |
| | **tube heading machine** [tju:b 'hedɪŋ məʃi:n] | Tubenverschlussautomat |
| | **waste suction unit** [weɪst 'sʌkʃn ju:nɪt] | Abfallabsaugvorrichtung |
| | **hold** [həʊld] | (fest)halten |
| | **suck** [sʌk] | saugen |
| | **cap** [kæp] | Verschluss, Deckel |
| | **weld** [weld] | schweißen |
| | **Ltd (Limited)** ['lɪmɪtɪd] | GmbH |
| | **delivery** [dɪ'lɪvəri] | (Aus-)Lieferung |
| | **confirm** [kən'fɜ:m] | bestätigen |
| | **arrangement** [ə'reɪndʒmənt] | Absprache, Vereinbarung, Termin |
| | **deliver** [dɪ'lɪvə] | (aus)liefern |
| | **commission** [kə'mɪʃn] | übergeben, in Betrieb nehmen |
| | **contractor** [kən'træktə] | Subunternehmer |
| | **foundation** [faʊn'deɪʃn] | Fundament |
| | **dry out** [ˌdraɪ 'aʊt] | (aus)trocknen |
| | **originally** [ə'rɪdʒənəli] | ursprünglich |
| 47 | **doable** ['du:əbl] | machbar |
| | **operational** [ˌɒpə'reɪʃənl] | einsatzfähig, in Betrieb |
| | **schedule** ['ʃedju:l] | (Zeit-)Plan |
| | **delay** [dɪ'leɪ] | Verspätung, Verzögerung |
| | **up and running** [ˌʌp ən 'rʌnɪŋ] | einsatzbereit, funktionstüchtig |
| | **be go** [bi 'gəʊ] | funktionieren, arbeiten |
| | **line sth up** [ˌlaɪn 'ʌp] | etw organisieren |
| | **forklift** ['fɔ:k lɪft] | Gabelstapler |
| | **crane** [kreɪn] | Kran |
| | **on board** [ɒn 'bɔ:d] | an Bord |
| | **be a case of** [bi ə 'keɪs əv] | um ... gehen |
| | **lower** ['ləʊə] | senken, herablassen |
| | **get in touch** [get ɪn 'tʌtʃ] | sich in Verbindung setzen, sich melden |
| 48 | **arrange for** [ə'reɪndʒ fə] | organisieren |
| | **start sth up** [ˌstɑ:t 'ʌp] | etw in Betrieb setzen |
| | **provide** [prə'vaɪd] | zur Verfügung stellen, erbringen |
| | **achievable** [ə'tʃi:vəbl] | machbar |
| | **involved (in)** [ɪn'vɒlvd ɪn] | beteiligt (an) |
| | **safety guideline** [ˌseɪfti 'gaɪdlaɪn] | Sicherheitsbestimmung |
| | **excerpt** ['eksɜ:pt] | Auszug |
| | **refit** [ˌri:'fɪt] | wieder einsetzen |
| | **preventable** [prɪ'ventəbl] | vermeidbar |
| | **relating to** [rɪ'leɪtɪŋ tə] | in Bezug auf, in Zusammenhang mit, betreffend |
| 49 | **electricity supply** [ɪlekˌtrɪsəti sə'plaɪ] | Netz-, Stromanschluss |
| | **construct** [kən'strʌkt] | bauen, konstruieren |
| | **equip** [ɪ'kwɪp] | ausrüsten, ausstatten |
| | **hazard** ['hæzəd] | Risiko, Gefahr |
| | **specific** [spə'sɪfɪk] | speziell |
| | **in force** [ɪn 'fɔ:s] | in Kraft |
| | **apply to** [ə'plaɪ tə] | gelten für |
| | **static** ['stætɪk] | statisch |
| | **as to** [əz tə] | so dass |
| | **build-up** ['bɪld ʌp] | Bildung |
| | **potential(ly)** [pə'tenʃəl, -i] | potenziell |
| | **electrostatic charge** [ɪˌlektrəʊ'stætɪk tʃɑ:dʒ] | elektrostatische Ladung |
| | **fit with** ['fɪt wɪð] | ausstatten mit |
| | **discharging system** [dɪs'tʃɑ:dʒɪŋ sɪstəm] | Entladungssystem |
| | **render** ['rendə] | machen |
| | **failing this** ['feɪlɪŋ ðɪs] | falls nicht |
| | **housing** ['haʊzɪŋ] | Verkleidung, Gehäuse |
| | **connection** [kə'nekʃn] | Verbindung, Anschluss |
| | **fluid** ['flu:ɪd] | fließend |
| | **owing to** ['əʊɪŋ tə] | wegen, aufgrund |
| | **pipe** [paɪp] | Rohr, Röhre |
| | **cable** ['keɪbl] | Kabel |
| | **connector** [kə'nektə] | Verbindung(sstück) |
| 50 | **action plan** ['ækʃn plæn] | Aktionsplan, Verfahren |
| | **assist** [ə'sɪst] | helfen |

| | | |
|---|---|---|
| | **take into consideration** [ˌteɪk ɪntə kənˌsɪdə'reɪʃn] | berücksichtigen |
| | **successful** [sək'sesfl] | erfolgreich |
| | **brief** [bri:f] | einweisen, instruieren |
| | **check with sb** ['tʃek wɪð] | sich bei jdm vergewissern |
| | **conformity** [kən'fɔ:məti] | Übereinstimmung |
| | **remind** [rɪ'maɪnd] | (daran) erinnern |
| | **accommodation** [əˌkɒmə'deɪʃn] | Unterkunft |
| | **on hand** [ɒn 'hænd] | vorhanden |
| | **in time** [ɪn 'taɪm] | rechtzeitig |
| | **overland** ['əʊvəlænd] | über Land |
| | **calendar month** ['kælɪndə mʌnθ] | Kalendermonat |
| | **fly on** ['flaɪ ɒn] | weiterfliegen |
| 51 | **transportation** [ˌtrænspɔ:'teɪʃn] | Transport, Beförderung |
| | **mount** [maʊnt] | montieren, einbauen |
| | **safety device** ['seɪfti dɪvaɪs] | Sicherheitsvorrichtung |
| | **trial run** ['traɪəl rʌn] | Probelauf |
| | **according to** [ə'kɔ:dɪŋ tə] | entsprechend |
| | **jerk** [dʒɜ:k] | Erschütterung, Ruck, Stoß |
| | **handling** ['hændlɪŋ] | Handhabung, Umgang |
| | **capacity** [kə'pæsəti] | Kapazität, Leistung |
| | **authorized** ['ɔ:θəraɪzd] | befugt, autorisiert |
| | **prevention** [prɪ'venʃn] | Verhütung |
| | **regulation** [ˌregju'leɪʃn] | Bestimmung, Verordnung, Vorschrift |
| | **observe** [əb'zɜ:v] | beachten |
| | **means of transport** [mi:nz əf 'trænspɔ:t] | Beförderungsmittel |
| | **withstand** [wɪð'stænd] | aushalten |
| | **secure** [sɪ'kjuə] | absichern, festmachen |
| | **lifting device** [ˌlɪftɪŋ dɪ'vaɪs] | Hebevorrichtung |
| | **centre of gravity** [ˌsentər əv 'grævəti] | Schwerpunkt |
| | **operating** ['ɒpəreɪtɪŋ] | Arbeits- |
| | **by means of** [baɪ 'mi:nz əv] | mit Hilfe von |
| | **take care** [teɪk 'keə] | aufpassen |
| | **bring down** [ˌbrɪŋ 'daʊn] | absetzen |
| | **endanger** [ɪn'deɪndʒə] | gefährden |
| | **electrical supply** [ɪˌlektrɪkl sə'plaɪ] | Netz-, Stromanschluss |
| | **skilled** [skɪld] | ausgebildet, Fach- |
| | **rotation** [rəʊ'teɪʃn] | Drehung |
| | **brief(ly)** ['bri:f, -li] | kurz |
| | **press** [pres] | drücken |
| | **inlet** ['ɪnlet] | Einlass, Zufluss |
| | **return line** [rɪ'tɜ:n laɪn] | Rückleitung |
| | **cooling** ['ku:lɪŋ] | Kühl- |
| | **works-side** ['wɜ:ks saɪd] | der Maschine zugewandt |
| | **cooling system** ['ku:lɪŋ sɪstəm] | Kühlsystem |
| | **mix up** [ˌmɪks 'ʌp] | verwechseln |
| | **pneumatics line** [nju:'mætɪks laɪn] | Druckluftleitung |
| | **compressed air** [kəm'prest eə] | Druckluft |
| | **pneumatic** [nju:'mætɪk] | pneumatisch |
| | **assembly instructions** [əˌsembli ɪn'strʌkʃnz] | Bedienungsanleitung |
| | **emergency stop** [ɪ'mɜ:dʒənsi stɒp] | Notschalter |
| | **assigned** [ə'saɪnd] | vorgesehen |
| | **pad** [pæd] | Polster, Schützer |
| | **safety measure** ['seɪfti meʒə] | Sicherheitsmaßnahme |
| | **effectiveness** [ɪ'fektɪvnəs] | Effektivität, Wirksamkeit |
| | **set** [set] | einsetzen, -stellen |
| | **relief valve** [rɪ'li:f vælv] | Entlastungsventil |
| | **hydraulic** [haɪ'drɔ:lɪk] | hydraulisch |
| | **wire rope** ['waɪə rəʊp] | Drahtseil |
| | **length** [leŋθ] | Länge |
| 52 | **go wrong** [ˌgəʊ 'rɒŋ] | schiefgehen |
| | **safety regulation** ['seɪfti ˌregju'leɪʃn] | Sicherheitsbestimmung |
| | **facility** [fə'sɪləti] | Vor-, Einrichtung |
| | **hold up** [ˌhəʊld 'ʌp] | aufhalten |
| 52t | **overdue** [ˌəʊvə'dju:] | überfällig |
| | **manage** ['mænɪdʒ] | (es) schaffen |
| | **get through to sb** [ˌget 'θru: tə] | jdn erreichen |
| | **mobile phone** [ˌməʊbaɪl 'fəʊn] | Mobiltelefon, Handy |
| | **motorway** ['məʊtəweɪ] | Autobahn |
| | **be caught up in** [bi ˌkɔ:t 'ʌp ɪn] | feststecken in |
| | **tailback** ['teɪlbæk] | Autoschlange, (Rück-)Stau |
| | **be way behind with sth** [bi ˌweɪ bɪ'haɪnd wɪð] | in Rückstand geraten |
| | **upset** [ˌʌp'set] | besorgt |
| | **get sth done** [get 'dʌn] | etw erledigen |
| 52 | **explanation** [ˌeksplə'neɪʃn] | Erklärung |
| | **replacement** [rɪ'pleɪsmənt] | Ersatz(teil) |
| | **offer an apology** [ˌɒfər ən ə'pɒlədʒi] | eine Entschuldigung aussprechen |
| | **calm down** [ˌkɑ:m 'daʊn] | beruhigen, beschwichtigen |
| | **reassure** [ˌri:ə'ʃʊə] | versichern |
| | **rectify** ['rektɪfaɪ] | richtigstellen, korrigieren |
| 53 | **oh dear** [əʊ 'dɪə] | ach du liebe Zeit! |
| | **suggestion** [sə'dʒestʃən] | Vorschlag |
| 53p | **oversee** [ˌəʊvə'si:] | beaufsichtigen |
| | **make sth up** [ˌmeɪk 'ʌp] | etw erfinden, sich etw ausdenken |
| | **tight** [taɪt] | eng, knapp |
| | **assure** [ə'ʃʊə] | zu-, versichern, versprechen |
| | **supplier** [sə'plaɪə] | Lieferant/in, Zulieferer |
| | **strike** [straɪk] | Streik |
| | **on schedule** [ɒn 'ʃedju:l] | nach Plan, pünktlich |
| | **replacement part** [rɪ'pleɪsmənt pɑ:t] | Ersatzteil |

| | | |
|---|---|---|
| 54 | **run into trouble** [ˌrʌn ɪntə 'trʌbl] | Schwierigkeiten Ärger bekommen |
| | **unauthorized** [ˌʌn'ɔ:θəraɪzd] | unbefugt, unberechtigt |
| | **wiring** ['waɪərɪŋ] | Verdrahtung, Verkabelung, Anschluss |
| | **nasty** ['nɑ:sti] | schlimm, unangenehm |
| | **tie one's hair back** [taɪ wʌnz ˌheə 'bæk] | das Haar hochbinden |
| | **clumsy** ['klʌmzi] | ungeschickt |
| | **reschedule** [ˌri:'ʃedju:l] | neu planen, ansetzen |
| | **tip over** [ˌtɪp 'əʊvə] | umkippen |
| | **chain reaction** [ˌtʃeɪn ri'ækʃn] | Kettenreaktion |
| | **insulation** [ˌɪnsju'leɪʃn] | Isolierung, Isolation |
| | **circuit breaker** ['sɜ:kɪt breɪkə] | Stromunterbrecher, Überspannungsschalter |
| 55 | **CEO (Chief Executive Officer)** [ˌsi: i: 'əʊ, ˌtʃi:f ɪg'zekjətɪv ɒfɪsə] | Geschäftsführer/in |
| | **background** ['bækgraʊnd] | Hintergrund, Herkunft |
| | **on time** [ɒn 'taɪm] | pünktlich, rechtzeitig |
| | **a number of** [ə 'nʌmbər əv] | eine Reihe von |
| | **failure** ['feɪljə] | Versagen, Fehler |
| | **on sb's part** [ɒn ˌsʌmbədiz 'pɑ:t] | auf jds Seite |
| | **due to** ['dju: tə] | wegen, aufgrund von |
| | **under normal circumstances** [ˌʌndə ˌnɔ:ml 'sɜ:kəmstənsɪz] | normalerweise |
| | **make up (the) time** [meɪk ˌʌp ðə 'taɪm] | die Zeit einholen |
| | **deadline** ['dedlaɪn] | (Liefer-)Termin, Frist |
| | **extreme** [ɪk'stri:m] | extrem |
| | **weather conditions** ['weðə kəndɪʃnz] | Wetterbedingungen |
| | **scheduled** ['ʃedju:ld] | vorgesehen, geplant, planmäßig |
| | **take responsibility for** [teɪk rɪˌspɒnsə'bɪləti fə] | (die) Verantwortung übernehmen für |
| | **share responsibility for** [ʃeə rɪˌspɒnsə'bɪləti fə] | Mitverantwortung übernehmen für |
| | **become aware** [bɪˌkʌm ə'weə] | sich bewusst werden |
| | **forecast** ['fɔ:kɑ:st] | vorhersagen |
| | **postpone** [pə'spəʊn] | verschieben |
| | **insist** [ɪn'sɪst] | (darauf) bestehen |
| | **fail to do** [ˌfeɪl tə 'du:] | nicht tun |
| | **rear** [rɪə] | Rückseite |
| | **keep clear** [ki:p 'klɪə] | freihalten |
| | **despite** [dɪ'spaɪt] | trotz |
| | **access route** ['ækses ru:t] | Zugang(sweg) |
| | **as a result** [əz ə rɪ'zʌlt] | deswegen |

## Unit 6

| | | |
|---|---|---|
| 56 | **pump** [pʌmp] | pumpen |
| 57 | **my pleasure** [maɪ 'pleʒə] | gern (geschehen) |
| | **state** [steɪt] | feststellen, angeben |
| | **warranty** ['wɒrənti] | Garantie, Gewährleistung |
| | **thorough(ly)** ['θʌrə, -li] | gründlich |
| | **beforehand** [bɪ'fɔ:hænd] | vorher, zuvor |
| | **as a matter of fact** [əz ə ˌmætər əv 'fækt] | tatsächlich, in der Tat |
| | **at full efficiency** [ət ˌfʊl ɪ'fɪʃnsi] | mit voller Leistung(sfähigkeit) |
| | **inspection** [ɪn'spekʃn] | Kontrolle, Inspektion |
| | **slow down** [ˌsləʊ 'daʊn] | langsamer werden |
| | **at full speed** [ət ˌfʊl spi:d] | mit voller Geschwindigkeit |
| | **handbook** ['hændbʊk] | Bedienungsanleitung, Handbuch |
| | **sound as if** ['saʊnd əz ɪf] | klingen als ob |
| | **actuator** ['æktʃueɪtə] | Aktor |
| | **definitely** ['defɪnətli] | sicher(lich), auf jeden Fall |
| | **afford** [ə'fɔ:d] | (es) sich leisten (können) |
| | **downtime** ['daʊntaɪm] | Ausfallzeit, Produktionsleerlauf(zeit) |
| | **suit sb** [su:t] | jdm passen, jdm recht sein |
| | **I can't make it** [aɪ kɑ:nt 'meɪk ɪt] | es passt mir nicht |
| | **accompanied by** [ə'kʌmpənid baɪ] | begleitet von |
| | **splendid** ['splendɪd] | prima, ausgezeichnet |
| | **Bye for now** ['baɪ fə naʊ] | Tschüs dann |
| | **feedback** ['fi:dbæk] | Rückmeldung |
| | **urgent** ['ɜ:dʒənt] | dringend, eilig |
| 58 | **display** [dɪ'spleɪ] | Bildschirm, Anzeige |
| | **driven by** ['drɪvn baɪ] | angetrieben durch |
| | **written** ['rɪtn] | schriftlich |
| | **particular** [pə'tɪkjələ] | bestimmt |
| | **period** ['pɪəriəd] | (Zeit-)Raum |
| 59 | **greetings** ['gri:tɪŋz] | Grüße |
| | **honoured** ['ɒnəd] | geehrt |
| | **hold on** [ˌhəʊld 'ɒn] | dranbleiben |
| | **come along** [ˌkʌm ə'lɒŋ] | mitkommen |
| 59p | **repair job** [rɪ'peə dʒɒb] | Reparaturarbeit |
| | **seminar** ['semɪnɑ:] | Seminar, Fachtagung |
| | **customer relations** [ˌkʌstəmə rɪ'leɪʃnz] | Kundenbeziehungen |
| | **return flight** [rɪˌtɜ:n 'flaɪt] | Rückflug |
| 60 | **basic** ['beɪsɪk] | grundlegend, elementar |
| | **scientific** [ˌsaɪən'tɪfɪk] | wissenschaftlich |
| | **principle** ['prɪnsəpl] | Grundsatz, Prinzip |
| | **like** [laɪk] | gleich |
| | **pole** [pəʊl] | Pol |
| | **repel** [rɪ'pel] | (sich) abstoßen |
| | **unlike** [ˌʌn'laɪk] | ungleich |
| | **attract** [ə'trækt] | (sich) anziehen |
| | **electric current** [ɪˌlektrɪk 'kʌrənt] | (elektrischer) Strom |
| | **magnetic field** [mægˌnetɪk 'fi:ld] | Magnetfeld |
| | **essentially** [ɪ'senʃəli] | im Wesentlichen, im Grunde |
| | **rotate** [rəʊ'teɪt] | rotieren, (sich) drehen |

| Unit | English | German |
|---|---|---|
| | **electromagnet** [ɪ'lektrəʊmægnət] | Elektromagnet |
| | **transform into** [træns'fɔ:m ɪntə] | umwandeln in |
| | **DC motor** [ˌdi: 'si: məʊtə] | Gleichstrommotor |
| | **armature** ['ɑ:mətʃə] | Armatur; Anker |
| | **permanent** ['pɜ:mənənt] | ständig, permanent |
| | **opposing** [ə'pəʊzɪŋ] | gegenüber (liegend) |
| | **magnet** ['mægnət] | Magnet |
| | **exciter field** [ɪk'saɪtə fi:ld] | Erregerfeld |
| | **interaction** [ˌɪntər'ækʃn] | Interaktion |
| | **torque** [tɔ:k] | Drehmoment, -kraft |
| | **split-ring commutator** [ˌsplɪt rɪŋ 'kɒmjuteɪtə] | Stromwender mit Stromwenderlamellen |
| | **reverse** [rɪ'vɜ:s] | umkehren |
| | **magnetic** [mæg'netɪk] | magnetisch |
| | **brush** [brʌʃ] | (Kontakt-)Bürste, Stromabnehmer |
| | **brush against** ['brʌʃ əgenst] | streifen an |
| | **reversal** [rɪ'vɜ:sl] | Umkehrung |
| | **interact with** [ˌɪntər'ækt wɪð] | sich gegenseitig beeinflussen |
| | **alternately** [ɔ:l'tɜ:nətli] | abwechselnd |
| | **AC motor** [ˌeɪ 'si: məʊtə] | Wechselstrommotor |
| | **consist of** [kən'sɪst əv] | bestehen aus |
| | **stationary** ['steɪʃənri] | fest(stehend), Stand- |
| | **stator** ['steɪtə] | Ständer, feststehender Teil |
| | **coil** [kɔɪl] | Windung, Spule |
| | **supply with** [sə'plaɪ wɪð] | (be)liefern mit |
| | **rotor** ['rəʊtə] | Rotor, Läufer |
| | **shaft** [ʃɑ:ft] | Achse, (Haupt-)Welle |
| | **induction AC motor** [ɪnˌdʌkʃn ˌeɪ 'si: məʊtə] | Induktions-wechselstrommotor |
| | **produced by** [prə'dju:st baɪ] | erzeugt durch |
| | **connected by** [kə'nektɪd baɪ] | verbunden durch |
| | **solid** ['sɒlɪd] | fest |
| | **bar** [bɑ:] | Stab |
| | **induction motor** [ɪn'dʌkʃn məʊtə] | Induktionsmotor |
| | **synchronous AC motor** [ˌsɪnkrənəs ˌeɪ 'si: məʊtə] | Synchronwechselstrom-motor |
| | **speed** [spi:d] | Geschwindigkeit |
| | **alternator** ['ɔ:ltəneɪtə] | Wechselstromgenerator |
| | **synchronous motor** ['sɪnkrənəs məʊtə] | Synchronmotor |
| | **high-speed** ['haɪspi:d] | Hochgeschwindigkeits- |
| | **universal motor** [ˌju:nɪ'vɜ:sl məʊtə] | Universalmotor |
| | **supply current** [səˌplaɪ 'kʌrənt] | Stromzuleitung |
| | **polarity** [pə'lærəti] | Polarität |
| | **cycle** ['saɪkl] | Zyklus, Kreislauf |
| | **compact** ['kɒmpækt] | kompakt |
| | **exceed** [ɪk'si:d] | übersteigen, -treffen |
| | **revolution** [ˌrevə'lu:ʃn] | (Um-)Drehung |
| | **the mains current** [ðə 'meɪnz kʌrənt] | Netzstrom |
| | **appliance** [ə'plaɪəns] | Gerät |
| | **blender** ['blendə] | Mischer, Mixer |
| | **desire** [dɪ'zaɪə] | wünschen, verlangen |
| 61 | **rating plate** ['reɪtɪŋ pleɪt] | Typenschild |
| | **rated speed** ['reɪtɪd spi:d] | Nenndrehzahl |
| | **power factor** ['paʊə fæktə] | Leistungsfaktor |
| | **star connection** [stɑ: kə'nekʃn] | Sternschaltung |
| | **delta connection** [ˌdeltə kə'nekʃn] | Dreieckschaltung |
| | **mains frequency** [meɪnz 'fri:kwənsi] | Netzfrequenz |
| | **performance** [pə'fɔ:məns] | Leistung |
| | **rpm (revolutions per minute)** [ˌɑ: pi: 'em, ˌrevəˌlu:ʃnz pə 'mɪnɪt] | UpM (Umdrehungen pro Minute) |
| | **require** [rɪ'kwaɪə] | benötigen |
| 62 | **look over** [ˌlʊk 'əʊvə] | sich ansehen |
| | **control valve** [kən'trəʊl vælv] | Steuerventil |
| | **release** [rɪ'li:s] | Zufuhr |
| | **tank** [tæŋk] | Tank |
| | **regulator** ['regjuleɪtə] | Regler |
| | **pass through** ['pɑ:s θru:] | hindurchleiten |
| | **hold** [həʊld] | speichern, aufbewahren |
| | **buffer** ['bʌfə] | (Speicher-)Puffer |
| | **air pressure** ['eə preʃə] | Luftdruck |
| | **keep up** [ˌki:p 'ʌp] | aufrecht erhalten |
| | **at all times** [ət ɔ:l 'taɪmz] | jederzeit |
| 63 | **operating manual** [ˌɒpəreɪtɪŋ 'mænjuəl] | Handbuch, Betriebsanleitung |
| | **cabinet** ['kæbɪnət] | Kabinett, Schrank |
| | **clean out** [ˌkli:n 'aʊt] | reinigen, säubern |
| | **lubrication** [ˌlu:brɪ'keɪʃn] | Schmierung |
| | **Pos. (position)** [pə'zɪʃn] | Stelle, Position |
| | **nipple** ['nɪpl] | Nippel |
| | **valve** [vælv] | Ventil |
| | **filter cartridge** ['fɪltə kɑ:trɪdʒ] | Filterpatrone |
| | **fittings** ['fɪtɪŋz] | Anschlüsse |
| | **pass** [pɑ:s] | leiten |
| | **compress** [kəm'pres] | komprimieren, zusammenpressen |
| 64 | **accumulator** [ə'kju:mjəleɪtə] | Akkumulator |
| | **heat exchanger** [ˌhi:t ɪks'tʃeɪndʒə] | Wärmetauscher |
| | **oil level indicator** [ˌɔɪl levl 'ɪndɪkeɪtə] | Ölstandsanzeige |
| | **sensor/actuator box** [sensə 'æktʃueɪta bɒks] | Sensor-/Aktor-Box |
| | **thermostatic valve** [ˌθɜ:mə'stætɪk vælv] | Thermostat(ventil) |
| | **valve block** ['vælv blɒk] | Ventilkammer, -gehäuse |
| | **various** ['veəriəs] | verschiedene |
| 64t | **complicated** ['kɒmplɪkeɪtɪd] | kompliziert |

| | | |
|---|---|---|
| | **straightforward** [ˌstreɪt'fɔ:wəd] | einfach |
| | **dial** ['daɪəl] | (Wähl-)Scheibe |
| | **cylinder** ['sɪlɪndə] | Zylinder |
| | **indicator** ['ɪndɪkeɪtə] | Anzeige |
| | **have a go** [həv ə 'gəʊ] | es versuchen |
| | **leave sth to sb** ['li:v tə] | jdm etw überlassen |
| **64** | **visible** ['vɪzəbl] | sichtbar |
| | **located** [ləʊ'keɪtɪd] | gelegen |
| | **refer to** [rɪ'fɜ: tə] | sich beziehen auf |
| | **bladder** ['blædə] | Blase |
| | **nitrogen** ['naɪtrədʒən] | Stickstoff |
| | **mushroom valve** ['mʌʃrʊm vælv] | Tellerventil |
| | **expand** [ɪk'spænd] | (sich) ausdehnen |
| | **volume** ['vɒlju:m] | Volumen |
| **65** | **overhaul** ['əʊvəhɔ:l] | Überholung |
| | **plant** [plɑ:nt] | Maschinen, Anlage |
| | **churn out** [ˌtʃɜ:n 'aʊt] | (massenweise) herstellen |
| | **lifespan** ['laɪfspæn] | Lebenszeit, -dauer |
| | **scrap** [skræp] | verschrotten, entsorgen |
| | **disposable** [dɪ'spəʊzəbl] | Einweg-, Wegwerf- |
| | **matter** ['mætə] | Sache, Angelegenheit |
| | **hugely** ['hju:dʒli] | riesig, ungeheuer |
| | **continue** [kən'tɪnju:] | fortfahren |
| | **get a return on sth** [ˌget ə rɪ'tɜ:n ɒn] | einen Gewinn durch etw erzielen |
| | **investment** [ɪn'vestmənt] | Investition |
| | **lie behind** [ˌlaɪ bɪ'haɪnd] | stecken hinter |
| | **strategy** ['strætədʒi] | Strategie |
| | **MRO (Maintenance, Repair and Overhaul)** [ˌem ɑ:r 'əʊ, ˌmeɪntənəns rɪˌpeə ənd 'əʊvəhɔ:l] | Wartung, Reparatur und Überholung |
| | **increase** [ɪn'kri:s] | zunehmen, anwachsen, steigen |
| | **manufacturer** [ˌmænju'fæktʃərə] | Hersteller/in, Produzent/in |
| | **squeeze** [skwi:z] | herausholen |
| | **profitable** ['prɒfɪtəbl] | gewinnbringend, rentabel |
| | **capital** ['kæpɪtl] | Kapital |
| | **consultant** [kən'sʌltənt] | Berater/in |
| | **realize** ['rɪəlaɪz] | verwirklichen, umsetzen |
| | **popular** ['pɒpjələ] | beliebt, populär |
| | **turn over** [ˌtɜ:n 'əʊvə] | übertragen |
| | **requirement** [rɪ'kwaɪəmənt] | Bedingung, Anforderung |
| | **self-explanatory** [ˌself ɪk'splænətri] | unmittelbar verständlich |
| | **approach** [ə'prəʊtʃ] | Ansatz, Herangehen(sweise) |
| | **remanufacture** [ˌri:ˌmænju'fæktʃə] | (vor Ort) warten und reparieren |
| | **recertify** [ˌri:'sɜ:tɪfaɪ] | (im Werk) komplett überholen, neu aufbauen |
| | **on site** [ɒn 'saɪt] | vor Ort |
| | **take back** [ˌteɪk 'bæk] | zurücknehmen |
| | **saving** ['seɪvɪŋ] | Einsparung, Ersparnis |
| | **obvious** ['ɒbviəs] | offensichtlich |
| | **starting point** ['stɑ:tɪŋ pɔɪnt] | Ausgangspunkt |
| | **overhauled** [ˌəʊvə'hɔ:ld] | überholt |
| | **duplicate** ['dju:plɪkeɪt] | duplizieren |
| | **duplication** [ˌdju:plɪ'keɪʃn] | Verdoppelung |
| | **cut out** [ˌkʌt 'aʊt] | ausschließen |
| | **retraining** [ˌri:'treɪnɪŋ] | Umschulung |
| | **play a role in** [pleɪ ə 'rəʊl ɪn] | eine Rolle spielen bei |
| | **frontliner** [ˌfrʌnt'laɪnə] | Vorreiter/in, Vorkämpfer/in |
| | **dispose of** [dɪ'spəʊz əv] | beseitigen, entsorgen |
| | **throw away** [ˌθrəʊ ə'weɪ] | wegwerfen, entsorgen |
| | **invest** [ɪn'vest] | investieren, anlegen |
| | **wherever** [weər'evə] | wo (auch) immer |
| | **pass along** [ˌpɑ:s ə'lɒŋ] | transportiert werden |
| | **put together** [ˌpʊt tə'geðə] | zusammensetzen |

## Unit 7

| | | |
|---|---|---|
| **66** | **presentation** [ˌprezn'teɪʃn] | Vorführung, Vorstellung, Präsentation |
| | **audience** ['ɔ:diəns] | Publikum, Zuhörer, Besucher |
| | **informal(ly)** [ɪn'fɔ:məl, -i] | ungezwungen, nicht förmlich |
| | **slide** [slaɪd] | Dia(positiv), Folie |
| | **prepare** [prɪ'peə] | vorbereiten |
| | **disorganized** [dɪs'ɔ:gənaɪzd] | chaotisch |
| | **body language** ['bɒdi læŋgwɪdʒ] | Körpersprache |
| | **hand movement** ['hænd mu:vmənt] | Handbewegung |
| | **get wrong** [ˌget 'rɒŋ] | missverstehen |
| | **give a presentation** [gɪv ə ˌprezn'teɪʃn] | eine Präsentation durchführen |
| **67** | **apology** [ə'pɒlədʒi] | Entschuldigung |
| | **by the way** [baɪ ðə 'weɪ] | übrigens, nebenbei |
| | **comment on** ['kɒment ɒn] | kommentieren |
| | **capability** [ˌkeɪpə'bɪləti] | Fähigkeit |
| | **application** [ˌæplɪ'keɪʃn] | Verwendung(szweck), Einsatz |
| | **focus on** ['fəʊkəs ɒn] | sich konzentrieren auf |
| | **outline** ['aʊtlaɪn] | umreißen, skizzieren |
| | **expansion** [ɪk'spænʃn] | Erweiterung |
| | **module** ['mɒdju:l] | Modul |
| | **turn one's attention to** [tɜ:n wʌnz ə'tenʃn tə] | sich zuwenden |
| | **programming** ['prəʊgræmɪŋ] | Programmieren |
| | **compatibility** [kəmˌpætə'bɪləti] | Kompatibilität |
| | **troubleshooting** ['trʌblʃu:tɪŋ] | Fehlersuche |
| | **hands-on** [ˌhændz 'ɒn] | aktiv, (selbst) zupackend |
| | **get one's hands on sth** [ˌget wʌnz 'hændz ɒn] | mit etw arbeiten |
| | **topic** ['tɒpɪk] | Thema |
| | **problem-solving** ['prɒbləm sɒlvɪŋ] | Problemlösung |

| | | |
|---|---|---|
| 68 | **be good at sth** [bi 'gʊd ət] | gut in etw sein, etw gut können |
| | **prefer** [prɪ'fɜ:] | vorziehen, lieber mögen, bevorzugen |
| | **move on** [ˌmu:v 'ɒn] | fortfahren |
| | **towards the end** [təˌwɔ:dz ði 'end] | am Ende |
| 69 | **installation configuration** [ɪnstəˌleɪʃn kənˌfɪgə'reɪʃn] | Konfiguration, Montageanordnung |
| | **customer service** [ˌkʌstəmə 'sɜ:vɪs] | Kundendienst |
| | **CPU (Central Processing Unit)** [ˌsi: pi: 'ju:, ˌsentrəl 'prəʊsesɪŋ ju:nɪt] | Zentraleinheit, -gerät, -baugruppe, Prozessor |
| | **a variety of** [ə və'raɪəti əv] | eine Vielzahl von, verschiedene |
| | **optional** ['ɒpʃənl] | wahlweise, optional |
| | **discrete I/O point** [dɪˌskri:t ˌaɪ 'əʊ pɔɪnt] | digitaler Ein- und Ausgang |
| | **stand-alone** ['stænd əˌləʊn] | allein stehend |
| | **get started with** [get 'stɑ:tɪd wɪð] | anfangen mit |
| | **panel** ['pænl] | Platte, Schalttafel |
| | **rail** [reɪl] | Schiene |
| | **horizontal(ly)** [ˌhɒrɪ'zɒntəl, -i] | horizontal, waagerecht |
| | **I/O expansion cable** [ˌaɪ 'əʊ ɪk'spænʃn keɪbl] | Erweiterungsskabel für Ein- und Ausgänge |
| | **flexibility** [ˌfleksə'bɪləti] | Flexibilität |
| | **mounting configuration** [ˌmaʊntɪŋ kənˌfɪgə'reɪʃn] | Einbaukonfiguration, Montageanordnung |
| | **FAQs (Frequently Asked Questions)** [ˌef eɪ 'kju:z, ˌfri:kwəntli ɑ:skt 'kwestʃənz] | häufig gestellte Fragen |
| | **monitor** ['mɒnɪtə] | überwachen, kontrollieren |
| 70 | **bullet point** ['bʊlɪt pɔɪnt] | Stichpunkt |
| | **accommodate** [ə'kɒmədeɪt] | unterbringen |
| | **clearance** ['klɪərəns] | Zwischenraum, Spiel, Abstand |
| | **execute** ['eksɪkju:t] | ausführen |
| | **port** [pɔ:t] | Anschluss, Buchse, Schnittstelle |
| | **status** ['steɪtəs] | Status, Stand |
| | **expansion module** [ɪkˌspænʃn 'mɒdju:l] | Erweiterungsmodul |
| | **designed for** [dɪ'zaɪnd fə] | bestimmt für |
| | **convection** [kən'vekʃn] | Konvektion |
| | **base unit** ['beɪs ju:nɪt] | Grund-, Zentralgerät |
| | **additional** [ə'dɪʃənl] | zusätzlich |
| | **space** [speɪs] | Raum, Platz |
| | **engage** [ɪn'geɪdʒ] | einstellen, -schalten, einstecken |
| | **disengage** [ˌdɪsɪn'geɪdʒ] | ausschalten, ziehen |
| | **expansion connector** [ɪk'spænʃn kənektə] | Erweiterungsverbindung |
| | **visual** ['vɪʒuəl] | visuell, optisch |
| | **sort into** ['sɔ:t ɪntə] | sortieren nach |
| 71t | **chap** [tʃæp] | Typ |
| | **clarify** ['klærəfaɪ] | erläutern, deutlich machen |
| | **communication processor card** [kəˌmju:nɪˌkeɪʃn 'prəʊsesə kɑ:d] | Kommunikationsprozessor-karte |
| | **multipoint interface cable** [ˌmʌltipɔɪnt 'ɪntəfeɪs keɪbl] | Mehrfachschnittstellen-kabel |
| | **clear up** [ˌklɪər 'ʌp] | (auf)klären |
| | **bus expansion connector** [ˌbʌs ɪk'spænʃn kənektə] | Busverbinder |
| | **spot on** ['spɒt ɒn] | haargenau richtig |
| 71 | **presenter** [prɪ'zentə] | Moderator/in |
| | **member** ['membə] | Mitglied |
| 72 | **rearrange** [ˌri:ə'reɪndʒ] | (um)ändern |
| | **spoken** ['spəʊkən] | gesprochen, mündlich |
| | **clarification** [ˌklærɪfɪ'keɪʃn] | Erklärung, Klärung, Erläuterung(en) |
| | **think up** [ˌθɪŋk 'ʌp] | (sich) ausdenken |
| | **PLC unit** [ˌpi: el 'si: ju:nɪt] | PLC-Einheit |
| 73 | **go on with** [ˌəʊ 'ɒn wɪð] | fortfahren mit |
| | **operating cycle** ['ɒpəreɪtɪŋ saɪkl] | Ablaufprogramm |
| | **flow chart** ['fləʊ tʃɑ:t] | Flussdiagramm |
| | **activate** ['æktɪveɪt] | aktivieren, einschalten |
| | **execution** [ˌeksɪ'kju:ʃn] | Aus-, Durchführung |
| | **record** [rɪ'kɔ:d] | aufnehmen, -zeichnen |
| | **stage** [steɪdʒ] | Phase, Abschnitt, Stadium |
| | **constant(ly)** ['kɒnstənt, -li] | ständig |
| | **envisage as** [ɪn'vɪzɪdʒ əz] | sich vorstellen als |
| | **critical** ['krɪtɪkl] | kritisch, entscheidend |
| | **in reality** [ɪn ri'æləti] | in Wirklichkeit, tatsächlich |
| | **gain** [geɪn] | gewinnen, erlangen |
| | **general** ['dʒenrəl] | allgemein, generell |
| | **counter** ['kaʊntə] | Zähler |
| | **timer** ['taɪmə] | Zeitnehmer, -schalter |
| | **update** ['ʌpdeɪt] | neuester/letzter Stand, neueste Version |
| | **examine** [ɪg'zæmɪn] | untersuchen, prüfen |
| | **previous** ['pri:viəs] | vorhergehend, vorherig |
| | **on the basis** [ɒn ðə 'beɪsɪs] | auf der Basis, Grundlage |
| 74 | **STL instruction** [ˌes ti: 'el ɪn'strʌkʃn] | STL-Anleitung |
| | **input/output control** [ˌɪnpʊt, 'aʊtpʊt kəntrəʊl] | Eingangs- und Ausgangsregler |
| | **failed** [feɪld] | misslungen |
| | **in relation to** [ɪn rɪ'leɪʃn tə] | in Bezug auf, im Verhältnis zu |
| | **illustrate** ['ɪləstreɪt] | illustrieren, veranschaulichen |
| 75 | **standard rail** ['stændəd reɪl] | Hutschiene |
| | **disable** [dɪs'eɪbl] | abschalten, deaktivieren |
| | **personal injury** [ˌpɜ:sənl 'ɪndʒəri] | Körperverletzung, Personenschaden |

| | | |
|---|---|---|
| | **related** [rɪ'leɪtɪd] | (miteinander) verbunden, angeschlossen |
| | **mounting** ['maʊntɪŋ] | Montage-, Einbau- |
| | **snap sth open** [ˌsnæp 'əʊpən] | etw öffnen |
| | **clip** [klɪp] | Klammer, Halterung, Rasthaken |
| | **hook onto** [hʊk 'ɒntə] | einhängen, -haken |
| | **snap sth closed** [ˌsnæp 'kləʊzd] | etw schließen, einrasten lassen |
| | **secure(ly)** [sɪ'kjuə, -li] | sicher, fest |
| | **vibration** [vaɪ'breɪʃn] | Schwingung |
| | **bus expansion port** [ˌbʌs ɪk'spænʃn pɔ:t] | Buserweiterungsanschluss |
| | **cover** [ˌkʌvə] | Abdeckung |
| | **existing** [ɪg'zɪstɪŋ] | bestehend, vorhanden |
| | **insert** [ɪn'sɜ:t] | einsetzen, einführen |
| | **pry sth away** [ˌpraɪ ə'weɪ] | etw aufhebeln |
| | **connecting joint** [kəˌnektɪŋ 'dʒɔɪnt] | Verbindungsstück, Halterung |
| | **proceed** [prə'si:d] | (fort)fahren, vorgehen |
| | **caution** ['kɔ:ʃn] | Vorsicht, Umsicht |
| | **bus connector** [bʌs kə'nektə] | Busverbinder |
| | **snap into place** [ˌsnæp ɪntə 'pleɪs] | (richtig) einrasten |
| | **orient** ['ɔ:rient] | ausrichten |
| | **with respect to** [wɪð rɪ'spekt tə] | mit Blick auf |
| 76 | **OHP (overhead projector)** [ˌəɪ eɪtʃ 'pi:, ˌəʊvəhed prə'dʒektə] | Overheadprojektor |
| | **clip-on microphone** [ˌklɪp ɒn 'maɪkrəfəʊn] | Ansteckmikrophon |
| | **whiteboard** ['waɪtbɔ:d] | Weißwandtafel |
| | **marker pen** ['mɑ:kə pen] | Folien-/Tafelstift |
| | **signposting phrase** ['saɪnpəʊstɪŋ freɪz] | strukturierender Ausdruck |
| | **respective** [rɪ'spektɪv] | jeweilig |
| | **category** ['kætəgəri] | Kategorie, Klasse |
| | **summary** ['sʌməri] | Zusammenfassung |
| | **conclusion** [kən'klu:ʒn] | Ende, Abschluss |
| | **once again** [wʌns ə'gen] | noch einmal |
| | **detailed** ['di:teɪld] | detailliert, ausführlich, genau |
| | **review** [rɪ'vju:] | wiederholen, zusammenfassen |
| 77 | **ladder logic** ['lædə lɒdʒɪk] | Strickleiterlogik |
| | **schematic** [ski:'mætɪk] | schematisch(e Darstellung) |
| | **graphical** ['græfɪkl] | grafisch |
| | **invent** [ɪn'vent] | erfinden |
| | **relay** ['ri:leɪ] | Relais |
| | **observation** [ˌɒbzə'veɪʃn] | Beobachtung |
| | **resemble** [rɪ'zembl] | ähneln, gleichen |
| | **a series of** [ə 'sɪəri:z əv] | Serie, Reihe |
| | **rung** [rʌŋ] | Sprosse |
| | **a set of** [ə 'set əv] | eine Reihe von |
| | **similarity** [ˌsɪmə'lærəti] | Ähnlichkeit |
| | **widely used** [ˌwaɪdli 'ju:zd] | weit verbreitet |
| | **sequential** [sɪ'kwenʃl] | sequenziell, aufeinander folgend |
| | **required** [rɪ'kwaɪəd] | verlangt, erforderlich |
| | **rework** [ˌri:'wɜ:k] | überarbeiten |
| | **hardwired** [ˌhɑ:d'waɪəd] | verdrahtet |
| | **sophisticated** [sə'fɪstɪkeɪtɪd] | hoch entwickelt, differenziert, kompliziert |
| | **vendor** ['vendə] | Verkäufer/in |
| | **associated** [ə'səʊʃieɪtɪd] | dazugehörig, verbunden |
| | **compatible** [kəm'pætəbl] | kompatibel, zueinander passend |
| | **notation** [nəʊ'teɪʃn] | Aufzeichnung, Bezeichnung |
| | **seamless(ly)** ['si:mləs, -li] | nahtlos |
| | **interchange** [ˌɪntə'tʃeɪndʒ] | austauschen |
| | **keyword** ['ki:wɜ:d] | Schlüssel-, Stichwort |
| | **drawing** ['drɔ:ɪŋ] | Zeichnung |
| | **intricate** ['ɪntrɪkət] | kompliziert |
| | **with ease** [wɪð 'i:z] | mit Leichtigkeit |
| | **join up** [ˌdʒɔɪn 'ʌp] | sich anschließen (lassen) |
| | **experience** [ɪk'spɪəriəns] | (Berufs-)Erfahrung |

## Unit 8

| | | |
|---|---|---|
| 78 | **apply for** [ə'plaɪ fə] | sich bewerben um |
| | **promotional prospect** [prəˌməʊʃənl 'prɒspekt] | Berufsaussichten, -chancen |
| | **responsibility** [rɪˌspɒnsə'bɪləti] | Verantwortlichkeit |
| | **earn** [ɜ:n] | verdienen |
| | **on top of** [ɒn 'tɒp əv] | zusätzlich |
| | **membership** ['membəʃɪp] | Mitgliedschaft |
| | **factor** ['fæktə] | Faktor |
| | **rank** [ræŋk] | einstufen, anordnen |
| | **importance** [ɪm'pɔ:tns] | Wichtigkeit, Bedeutung |
| | **care about** ['keər əbaʊt] | Wert legen auf |
| | **advert (advertisement)** ['ædvɜ:t, əd'vɜ:tɪsmənt] | Anzeige |
| | **newly qualified** [ˌnju:li 'kwɒlɪfaɪd] | frisch qualifiziert, ausgebildet |
| | **candidate** ['kændɪdət] | Bewerber/in |
| | **policy** ['pɒləsi] | (Firmen-)Politik |
| | **ability** [ə'bɪləti] | Fähigkeit, Können |
| | **opportunity** [ˌɒpə'tju:nəti] | Chance, Möglichkeit |
| | **auto mechanic** [ˌɔ:təʊ mɪ'kænɪk] | Kfz-Mechaniker/in |
| | **electronics engineer** [ɪlekˌtrɒnɪks endʒɪ'nɪə] | Elektroniker/in |
| | **electromechanical technician** [ɪˌlektrəʊmɪˌkænɪkl tek'nɪʃn] | Elektromechaniker/in |
| | **sensor technician** [ˌsensə tek'nɪʃn] | Messtechniker/in |
| | **laser technician** [ˌleɪzə tek'nɪʃn] | Lasertechniker/in |
| | **robotics engineer** [rəʊˌbɒtɪks endʒɪ'nɪə] | Roboteringenieur/in |
| 79 | **in the region of** [ɪn ðə 'ri:dʒən əv] | im Bereich von |
| | **fast-growing** [ˌfɑ:st 'grəʊɪŋ] | rasch wachsend |

| | | |
|---|---|---|
| | **specialize in** ['speʃəlaɪz ɪn] | sich spezialisieren auf |
| | **scanning equipment** [ˌskænɪŋ ɪ'kwɪpmənt] | Scanner(geräte) |
| | **motivated** ['məʊtɪveɪtɪd] | motiviert |
| | **support** [sə'pɔ:t] | verstärken, unterstützen |
| | **on your own initiative** [ɒn jɔ:r ˌəʊn ɪ'nɪʃətɪv] | selbständig |
| | **team player** ['ti:m pleɪə] | Mannschaftsspieler/in |
| | **think on one's feet** [ˌθɪŋk ɒn wʌnz 'fi:t] | schnell reagieren |
| | **effective(ly)** [ɪ'fektɪv, -li] | effektiv, wirksam, wirkungsvoll |
| | **additionally** [ə'dɪʃənəli] | darüber hinaus |
| | **have an eye for detail** [həv ən ˌaɪ fə 'di:teɪl] | Details wahrnehmen |
| | **communicator** [kə'mju:nɪkeɪtə] | Kommunikator/in |
| | **liaise with** [li'eɪz wɪð] | sich in Verbindung setzen mit |
| | **manufacturing equipment** [ˌmænjuˌfæktʃərɪŋ ɪ'kwɪpmənt] | Produktionsmaschinen |
| | **accredited** [ə'kredɪtɪd] | zugelassen, anerkannt |
| | **work experience** ['wɜ:k ɪkspɪəriəns] | Berufs-, Arbeitserfahrung |
| | **demonstrate** ['demənstreɪt] | zeigen, vorstellen, vorführen |
| | **understanding** [ˌʌndə'stændɪŋ] | Verständnis, Kenntnis(se) |
| | **sensor application** [ˌsensə ˌæplɪ'keɪʃn] | Sensoreinsatz |
| | **sensor system** [ˌsensə 'sɪstəm] | Sensorsystem |
| | **driving licence** ['draɪvɪŋ laɪsns] | Führerschein |
| | **closing date** ['kləʊzɪŋ deɪt] | Einsendeschluss |
| | **CV (curriculum vitae)** [kəˌrɪkjələm 'vi:taɪ] | Lebenslauf |
| | **covering letter** [ˌkʌvərɪŋ 'letə] | Begleitbrief, Anschreiben |
| | **referee** [ˌrefə'ri:] | Referenz(geber) |
| | **assistant** [ə'sɪstənt] | Assistent/in |
| | **equal opportunities** [ˌi:kwəl ˌɒpə'tju:nətiz] | Gleichberechtigung |
| | **select** [sɪ'lekt] | auswählen |
| | **regardless of** [rɪ'gɑ:dləs əv] | ungeachtet, ohne Rücksicht auf |
| | **gender** ['dʒendə] | Geschlecht |
| | **ethnicity** [eθ'nɪsəti] | ethnische Zugehörigkeit |
| | **sexual orientation** [ˌsekʃuəl ˌɔ:riən'teɪʃn] | sexuelle Orientierung |
| | **personal quality** [ˌpɜ:sənl 'kwɒləti] | persönliche Eigenschaft |
| 80 | **suited to** ['su:tɪd tə] | geeignet für |
| | **equivalent** [ɪ'kwɪvələnt] | Entsprechung, Übersetzung |
| | **vocational college/school** [vəʊˌkeɪʃənl 'kɒlɪdʒ, 'sku:l] | Berufsschule, -kolleg, berufsbildende Schule |
| | **primary (GB) / elementary (US) school** ['praɪməri, ˌelɪ'mentri sku:l] | Grundschule |
| | **GCSE (General Certificate of Secondary Education)** [ˌdʒi: si: es 'i:, ˌdʒenrəl səˌtɪfɪkət əv ˌsekəndri edʒu'keɪʃn] | mittlerer Schulabschluss |
| | **A level (GB) / High School Diploma (US)** ['eɪ levl, ˌhaɪ sku:l dɪ'pləʊmə] | (entspricht) Abitur |
| | **entrance qualification** [ˌentrəns ˌkwɒlɪfɪ'keɪʃn] | Reife |
| | **university of applied science** [ˌju:nɪ'vɜ:səti əv əˌplaɪd 'saɪəns] | Fachhochschule |
| | **apprenticeship scheme** [ə'prentɪʃɪp ski:m] | Duales System |
| | **comprehensive school (GB)** [ˌkɒmprɪ'hensɪv sku:l] | Gesamtschule |
| | **secondary (GB) / high (US) school** ['sekəndri, 'haɪ sku:l] | weiterführende Schule, Haupt-, Realschule |
| | **vocational training** [vəʊˌkeɪʃənl 'treɪnɪŋ] | (Berufs-)Ausbildung |
| | **grammar (GB) / high (US) school** ['græmə, 'haɪ sku:l] | Gymnasium |
| 81 | **certificate** [sə'tɪfɪkət] | Zeugnis, Urkunde |
| | **reference** ['refrəns] | Referenz, Empfehlungsschreiben |
| | **at a later date** [ət ə ˌleɪtə 'deɪt] | zu einem späteren Zeitpunkt |
| | **contact detail** [ˌkɒntækt 'di:teɪl] | Kontakt(adresse) |
| | **unless** [ən'les] | wenn nicht, es sei denn |
| | **case** [keɪs] | Fall |
| | **employment** [ɪm'plɔɪmənt] | Beschäftigung, Anstellung |
| | **law** [lɔ:] | Gesetz |
| | **racial discrimination** [ˌreɪʃl dɪˌskrɪmɪ'neɪʃn] | Rassendiskriminierung |
| | **personal detail** [ˌpɜ:sənl 'di:teɪl] | persönliche Angabe |
| | **in reverse order** [ɪn rɪˌvɜ:s 'ɔ:də] | in umgekehrter Reihenfolge |
| | **work one's way back** [ˌwɜ:k wʌnz weɪ 'bæk] | sich rückwärts durcharbeiten |
| | **personal** ['pɜ:sənl] | persönlich, privat |
| | **language skills** ['læŋgwɪdʒ skɪlz] | Sprachkenntnisse |
| | **pastime** ['pɑ:staɪm] | Freizeitaktivität |
| | **application form** [ˌæplɪ'keɪʃn fɔ:m] | Bewerbungsformular |
| | **enclose** [ɪn'kləʊz] | beilegen, beifügen |
| | **classmate** ['klɑ:smeɪt] | Klassenkamerad/in |
| | **youth** [ju:θ] | Jugend |
| | **summer camp** ['sʌmə kæmp] | Sommer(ferien)lager |

| | | |
|---|---|---|
| | **helper** ['helpə] | Helfer/in, Hilfskraft |
| | **EU (European Union)** [ˌjʊərəpi:ən 'ju:niən] | Europäische Union |
| | **marital status** [ˌmærɪtl 'steɪtəs] | Familienstand |
| | **single** ['sɪŋgl] | alleinstehend, unverheiratet |
| | **inline skating** ['ɪnlaɪn skeɪtɪŋ] | Inlineskaten |
| | **nationality** [ˌnæʃə'næləti] | Staatsangehörigkeit |
| | **placement** ['pleɪsmənt] | Praktikum |
| | **adult education college** [ˌædʌlt edʒu'keɪʃn kɒlɪdʒ] | Volkshochschule |
| | **rhetoric** ['retərɪk] | Rhetorik |
| 82 | **enclosed** [ɪn'kləʊzd] | Anlage |
| | **hardworking** [ˌhɑ:d'wɜ:kɪŋ] | fleißig |
| | **moreover** [mɔ:r'əʊvə] | außerdem, zudem |
| | **extensive** [ɪk'stensɪv] | umfassend, ausgedehnt |
| | **Yours sincerely** [jɔ:z sɪn'sɪəli] | Mit freundlichen Grüßen |
| | **highly** ['haɪli] | äußerst, hoch |
| | **consider** [kən'sɪdə] | berücksichtigen, in Frage kommen |
| | **hesitate** ['hezɪteɪt] | zögern |
| | **career** [kə'rɪə] | (beruflicher) Werdegang, Beruf |
| | **advance** [əd'vɑ:ns] | Fortschritte machen, vorwärtskommen |
| | **learner** ['lɜ:nə] | Lerner/in |
| 83 | **notes** [nəʊts] | Notizen |
| | **recently** ['ri:sntli] | kürzlich, vor kurzem |
| | **willing to** ['wɪlɪŋ tə] | gewillt (sein), bereit(willig) zu |
| | **HR (Human Resources) Co-ordinator** [ˌeɪtʃ 'ɑ:, ˌhju:mən rɪ'sɔ:sɪz kəʊ 'ɔ:dɪneɪtə] | Koordinator/in in der Personalabteilung |
| | **interview** ['ɪntəvju:] | Vorstellungsgespräch |
| 83t | **final** ['faɪnl] | Abschluss- |
| | **make up one's mind** [meɪk ˌʌp wʌnz 'maɪnd] | sich entschließen |
| | **letter of application** [ˌletər əv æplɪ'keɪʃn] | Bewerbungsschreiben |
| | **be about to** [bi ə'baʊt tə] | im Begriff sein, kurz bevorstehen |
| | **key focus** ['ki: fəʊkəs] | Hauptschwerpunkt |
| | **analyse** ['ænəlaɪz] | analysieren |
| | **on an ongoing basis** [ɒn ən 'ɒngəʊɪŋ beɪsɪs] | (fort)laufend, andauernd |
| | **in the meantime** [ɪn ðə 'mi:ntaɪm] | in der Zwischenzeit |
| | **weakness** ['wi:knəs] | Schwäche |
| | **grasp** [grɑ:sp] | erfassen, begreifen, verstehen |
| | **spontaneous** [spɒn'teɪniəs] | spontan |
| | **challenging** ['tʃælɪndʒɪŋ] | herausfordernd |
| | **foreign** ['fɒrən] | ausländisch |
| | **get sth off to a good start** [ˌget 'ɒf tu ə gʊd 'stɑ:t] | einen guten Start für etw finden |
| 83 | **take notes** [teɪk 'nəʊts] | (sich) Notizen machen |
| | **monitoring** ['mɒnɪtərɪŋ] | Überwachung, Kontrolle |
| | **throughout** [θru:'aʊt] | überall (in), in ganz |
| | **surprising(ly)** [sə'praɪzɪŋ, -li] | überraschend(erweise) |
| | **interviewee** [ˌɪntəvju:'i:] | Gesprächs-, Interviewpartner/in |
| | **perfectionist** [pə'fekʃənɪst] | Perfektionist/in |
| | **even though** ['i:vn ðəʊ] | obwohl ja, selbst wenn |
| | **perfectionism** [pə'fekʃənɪzəm] | Perfektionismus |
| | **trait** [treɪt] | Eigenschaft, Charakterzug, Merkmal |
| 84 | **impressed** [ɪm'prest] | beeindruckt |
| | **take on** [ˌteɪk 'ɒn] | an-, übernehmen, einstellen |
| | **in business** [ɪn 'bɪznəs] | im Geschäft |
| 85 | **top marks** ['tɒp mɑ:ks] | Höchstnoten |
| | **sensor head** ['sensə hed] | Sensorkopf |
| | **switching unit** ['swɪtʃɪŋ ju:nɪt] | (Um-)Schaltgerät |
| | **up to** ['ʌp tə] | bis zu |
| | **emit** [ɪ'mɪt] | aussenden |
| | **light beam** ['laɪt bi:m] | Lichtstrahl |
| | **receiver** [rɪ'si:və] | Empfänger |
| | **limit switch** ['lɪmɪt swɪtʃ] | Endlagenschalter |
| | **pneumatic cylinder** [nju:ˌmætɪk 'sɪlɪndə] | Druckzylinder |
| | **change-over contact** [ˌtʃeɪndʒ əʊvə 'kɒntækt] | Umschalter, Polwender, Wechselkontakt |
| | **piston** ['pɪstən] | Kolben |
| | **inductive** [ɪn'dʌktɪv] | induktiv |
| | **proximity** [prɒk'sɪməti] | Nähe |
| | **capacitive** [kə'pæsɪtɪv] | kapazitiv |
| | **optoelectrical** [ˌɒptəʊɪ'lektrɪkl] | optoelektrisch |
| | **temp. (temperature)** ['temprətʃə] | Temperatur, Wärme- |
| | **dirt** [dɜ:t] | Schmutz |
| | **humid** ['hju:mɪd] | feucht |
| 85p | **interviewer** ['ɪntəvju:ə] | Leiter/in des Vorstellungsgesprächs, Interviewer/in |
| | **technological** [ˌteknə'lɒdʒɪkl] | technologisch |
| | **conclude** [kən'klu:d] | (ab)schließen |
| 86 | **teamwork** ['ti:mwɜ:k] | Gruppenarbeit |
| | **specification sheet** [ˌspesɪfɪ'keɪʃn ʃi:t] | Spezifikationsblatt |
| | **light barrier** ['laɪt bæriə] | Lichtschranke |
| | **designation** [ˌdezɪg'neɪʃn] | Bezeichnung |
| | **adjustment** [ə'dʒʌstmənt] | Einstellung |

| | | |
|---|---|---|
| | **reflection angle** [rɪ'flekʃn æŋgl] | Abstrahlwinkel |
| | **status display** [ˌsteɪtəs dɪ'spleɪ] | Statusanzeige |
| | **transmitter** [træns'mɪtə] | Sender |
| | **no-load running** [ˌnəʊ 'ləʊd rʌnɪŋ] | Stromaufnahme im Leerlauf |
| | **pollution** [pə'lu:ʃn] | Verschmutzung |
| | **frequency** ['fri:kwənsi] | Frequenz |
| | **ambient** ['æmbiənt] | umgebend |
| | **protective** [prə'tektɪv] | Umgebungs- |
| | **impact proof** ['ɪmpækt pru:f] | stoß-, schlagfest |
| | **sensitivity setter** [ˌsensə'tɪvəti setə] | Empfindlichkeitseinsteller |
| | **infra red** [ˌɪnfrə 'red] | infrarot |
| | **pulsed** [pʌlst] | gepulst |
| | **adjustable** [ə'dʒʌstəbl] | ein-, verstellbar |
| | **residual ripple** [rɪ'zɪdjuəl 'rɪpl] | Restwelligkeit |
| | **unit plug** ['ju:nɪt plʌg] | Gerätestecker |
| | **rotating** [rəʊ'teɪtɪŋ] | drehbar |
| | **flashing mode** ['flæʃɪŋ məʊd] | Blinkmodus |
| | **contaminated** [kən'tæmɪneɪtɪd] | verunreinigt, verschmutzt |
| | **misaligned** [ˌmɪsə'laɪnd] | Dejustage, falsch justiert |
| | **distance** ['dɪstəns] | Entfernung, Abstand |
| | **go about sth** [ˌgəʊ ə'baʊt] | in einer Sache vorgehen |
| | **process of elimination** [ˌprəʊses əv ɪˌlɪmɪ'neɪʃn] | Ausschlussverfahren |
| **87** | **insight** ['ɪnsaɪt] | Einblick |
| | **invaluable** [ɪn'væljuəbl] | unschätzbar, (sehr)wertvoll |
| | **step by step** [ˌstep baɪ 'step] | Schritt für Schritt |
| | **jobsite** ['dʒɒbsaɪt] | Internetseite mit Stellenanzeigen |
| | **check out** [ˌtʃek 'aʊt] | durchsehen, überprüfen |
| | **in question** [ɪn 'kwestʃən] | in Frage kommend, betreffend |
| | **work permit** [ˌwɜ:k 'pɜ:mɪt] | Arbeitserlaubnis |
| | **health insurance** ['helθ ɪnʃʊərəns] | Krankenversicherung |
| | **pension cover** ['penʃn kʌvə] | Rentenanspruch, -zahlung |
| | **immigration service** [ˌɪmɪ'greɪʃn sɜ:vɪs] | Einwanderungsstelle |
| | **research** [rɪ'sɜ:tʃ] | Nachforschung(en) |
| | **realistic(ally)** [ˌri:ə'lɪstɪk, -kli] | realistisch(erweise) |
| | **embassy** ['embəsi] | Botschaft |
| | **consulate** ['kɒnsjələt] | Konsulat |
| | **attitude towards** ['ætɪtju:d təwɔ:dz] | Einstellung zu |

# A–Z word list

Dieses Wörterverzeichnis enthält alle Wörter aus **Mechatronics Matters** in alphabetischer Reihenfolge. Nicht aufgeführt sind die Wörter aus der Liste des Grundwortschatzes. t = das Wort befindet sich in den *Transcripts* (Hörverständnisübungen). p = das Wort befindet sich in den *Pairwork Files*.

**A level (GB) 80** (entspricht) Abitur
**ability 78** Fähigkeit, Können
**be about to 83t** im Begriff sein, kurz bevorstehen
**absence 26** Abwesenheit
**AC motor 60** Wechselstrommotor
**acceptable 45** annehmbar, akzeptabel
**access 14** Zugang
**access route 55** Zugang(sweg)
**accessory 28p** Zubehör
**accident 15** Unfall
**accommodate 70** unterbringen
**accommodation 50** Unterkunft
**accompanied by 57** begleitet von
**according to 51** entsprechend
**accounts 11t** Buchhaltung, Rechnungswesen
**accredited 79** zugelassen, anerkannt
**accumulator 64t** Akkumulator
**accurate 30** genau, richtig
**achievable 48** machbar
**achieve 35** erreichen, erlangen, erzielen
**acquaint oneself with** 15 sich vertraut machen mit
**act like 36** arbeiten wie
**in action 17** in Aktion
**action plan 50** Aktionsplan, Verfahren
**activate 73** aktivieren, einschalten
**actually 45** wirklich, tatsächlich, eigentlich
**actuator 57** Aktor
**adapt 35** anpassen, adaptieren
**adapted from 32** adaptiert von
**in addition to this 25** darüber hinaus
**additional 70** zusätzlich
**additionally 79** darüber hinaus
**adjustable 86** ein-, verstellbar
**adjustment 86** Einstellung
**admin (administration) 16** Verwaltung
**administer 15** (durch)führen, anwenden
**administration block 19** Verwaltung(sgebäude)
**adult education college 81** Volkshochschule
**advance 82** Fortschritte machen, vorwärtskommen
**advert (advertisement) 78** Anzeige
**advertising 11t** Werbung, Reklame
**advice 29** Rat, Ratschlag
**afford 57** (es) sich leisten (können)
**after-sales service 11t** Kundendienst, Wartung
**be aimed at 35** (ab)zielen auf, anstreben
**air pressure 62** Luftdruck
**all over the world 41t** in der ganzen Welt
**Allen key 22** Inbusschlüssel, Innensechskantschlüssel
**allow 15** gewähren, ermöglichen, (zu)lassen
**along with 35** zusammen mit
**ALT Triggering Function 28p** ALT-Auslösefunktion
**alternately 60** abwechselnd
**alternating 36** abwechselnd
**alternative 21** Alternative, Auswahlantwort
**alternator 60** Wechselstromgenerator
**ambient 86** umgebend
**ambulance 15** Krankenwagen
**ammeter 36** Amperemeter
**amount 32** Menge
**amp (ampere) 32** Ampere
**amplifier 36** Verstärker
**analyse 83t** analysieren
**Anglo-American 11** angloamerikanisch
**antiseptic 15** antiseptisch
**apologize 28** sich entschuldigen
**(offer an) apology 52** Entschuldigung (eine ~ aussprechen)
**apology 67** Entschuldigung
**appliance 60** Gerät
**application 67** Verwendung(szweck), Einsatz
**application form 81** Bewerbungsformular
**apply for 78** sich bewerben um
**apply sth to sth 15** etw auf etw ausüben, anwenden
**apply to 49** gelten für
**appointment 6** Termin, Verabredung
**apprentice 8** Lehrling, Auszubildende/r
**apprenticeship 16** Lehre, Ausbildung
**apprenticeship scheme 80** Duales System
**approach 65** Ansatz, Herangehen(sweise)
**appropriate 21** angemessen, passend
**approx. (approximate) 28p** ungefähr, etwa, circa
**area 14** Bereich
**arithmetic 31** Rechnen, Arithmetik
**armature 60** Armatur; Anker
**arrange 43** arrangieren, anordnen
**arrange for 48** organisieren
**arrangement 46** Absprache, Vereinbarung, Termin
**as 19** während
**as a matter of fact 57** tatsächlich, in der Tat
**as a result 55** deswegen
**as far as ... is concerned 38** was ... anbetrifft
**as instructed 15** wie angegeben
**as to 49** so dass
**ask advice about 26** um Rat fragen wegen
**assembly 39** Montage
**assembly instructions 51** Bedienungsanleitung
**assembly line 39** Montageband
**assigned 51** vorgesehen
**assist 50** helfen
**assistance 15** Hilfe(stellung)
**assistant 79** Assistent/in
**associated 77** dazugehörig, verbunden
**assure 53p** zu-, versichern, versprechen

**at a later date 81** zu einem späteren Zeitpunkt
**at all times 62** jederzeit
**at full efficiency 57** mit voller Leistung(sfähigkeit)
**at full speed 57** mit voller Geschwindigkeit
**at the far end 17** am anderen Ende
**at the heart of 41** im Zentrum von
**atom 41** Atom
**attach to 44** befestigen an, verbinden mit
**attention 14** Aufmerksamkeit
**attitude towards 87** Einstellung zu
**attract 60** (sich) anziehen
**audible continuity 27** hörbarer Durchgang
**audience 66** Publikum, Zuhörer, Besucher
**authorize 14** autorisieren, genehmigen
**authorized 51** befugt, autorisiert
**auto mechanic 78** Kfz-Mechaniker/in
**automated 35** automatisiert
**automatic(ally) 35** automatisch
**automobile 35** Auto(mobil)
**automobile manufacturing industry 35** Autoindustrie
**auto-ranging 27** Selbsterstreckung
**available 28** (am Telefon) zu sprechen
**avoid 35** (ver)meiden
**have an awareness of sth 45** sich einer Sache bewusst sein
**background 55** Hintergrund, Herkunft
**balance 35** Gleichgewicht, Ausgleich
**band 9** Band
**bandage 15** Verband, Binde
**bandwidth 28p** Bandbreite
**bar 60** Stab
**base unit 70** Grund-, Zentralgerät
**be based 7** liegen, angesiedelt sein
**based on 40** auf der Grundlage von
**basement 19** Untergeschoss, Keller(geschoss)
**basic 60** grundlegend, elementar
**basically 9** im Grunde, eigentlich
**on a daily basis 35** täglich
**on an ongoing basis 83t** (fort)laufend, andauernd
**on the basis 73** auf der Basis, Grundlage
**battery 27** Batterie
**battery life 27** Batterielebensdauer
**become aware 55** sich bewusst werden
**beforehand 57** vorher, zuvor
**behaviour 41** Verhalten
**benefit 13** zusätzliche Leistung
**bet 37** wetten
**bi-directional 36** Zweirichtungs-
**bladder 64** Blase
**blanket 15** (Woll-)Decke
**bleeding 15** Blutung
**blender 60** Mischer, Mixer
**blister 15** Blase
**blob 43** Punkt
**block 14** blockieren, versperren
**blow (out) 24** (aus)blasen
**body language 66** Körpersprache
**body trauma 15** schwere Körperverletzung
**bolt 23** Bolzen, Schraube
**bother sb 37** jdn stören
**bracket 13** Klammer
**branch 8** Filiale, Zweigstelle, Niederlassung
**breadth 25** Breite
**break down 34t** durchbrechen
**breathe 15** atmen
**breathing 15** Atmung
**bridge rectifier 42** Brückengleichrichter
**brief 50** einweisen, instruieren
**brief(ly) 51** kurz
**bring down 51** absetzen
**broad 27** breit
**brush 60** (Kontakt-)Bürste, Stromabnehmer
**brush against 60** streifen an
**buffer 62** (Speicher-)Puffer
**build-up 49** Bildung
**bullet point 70** Stichpunkt
**burn 15** Verbrennung
**bus connector 75** Busverbinder
**bus expansion connector 71t** Busverbinder
**bus expansion port 75** Buserweiterungsanschluss
**in business 84** im Geschäft
**by means of 51** mit Hilfe von
**by oneself 21** allein
**by the way 67** übrigens, nebenbei
**Bye for now 57** Tschüs dann
**cabinet 63** Kabinett, Schrank
**cable 49** Kabel
**calculation 31** Rechenart, -operation
**calendar month 50** Kalendermonat
**calibrator 28p** Kalibrierer, Kalibrator
**calm down 52** beruhigen, beschwichtigen
**candidate 78** Bewerber/in
**cap 46** Verschluss, Deckel
**capability 67** Fähigkeit
**capacitance 27** Kapazität
**capacitance meter 26** Kapazitätsmesser
**capacitive 85** kapazitiv
**capacitor 26** Kondensator
**capacity 51** Kapazität, Leistung
**capital 65** Kapital
**car owner 35** Autobesitzer/in
**carbon 36** Kohlenstoff
**cardiopulmonary resuscitation 15** Herzmassage, künstliche Beatmung
**care about 78** Wert legen auf
**career 82** (beruflicher) Werdegang, Beruf
**carry out 14** aus-, durchführen
**case 81** Fall
**be a case of 47** um ... gehen
**catalogue 42** Katalog
**catch on 14** erfasst werden von, sich verfangen in
**category 76** Kategorie, Klasse
**be caught up in 52t** feststecken in
**cause 14** verursachen
**caution 75** Vorsicht, Umsicht
**cell 45** Zelle
**centre of gravity 51** Schwerpunkt
**CEO (Chief Executive Officer) 55** Geschäftsführer/in
**certificate 81** Zeugnis, Urkunde
**chain reaction 54** Kettenreaktion
**challenge 45** An-, Herausforderung, Aufgabe
**challenging 83t** herausfordernd
**chance 20t** Chance, Gelegenheit, Möglichkeit
**change-over contact 85** Umschalter, Polwender, Wechselkontakt

**changing room** **18** Umkleide(raum)
**chap** **71t** Typ
**characteristic curve** **34** Kennlinie
**in charge of** **11t** verantwortlich für
**chart** **12** Diagramm, Tabelle
**check for** **24** nachprüfen
**check out** **87** durchsehen, überprüfen
**check with sb** **50** sich bei jdm vergewissern
**choice** **35** Wahl, Auswahl
**churn out** **65** (massenweise) herstellen
**circuit board** **37** Platine, Schalttafel
**circuit breaker** **54** Stromunterbrecher, Überspannungsschalter
**circuit diagram** **43** Schaltbild
**circuit** **26** Schaltung, (Strom-)Kreis
**circuitry** **26** Schaltung, Kreislauf, (Strom-)Kreis
**clarification** **72** Erklärung, Klärung, Erläuterung(en)
**clarify** **71t** erläutern, deutlich machen
**classmate** **81** Klassenkamerad/in
**clean out** **63** reinigen, säubern
**clean up** **14** säubern, aufwischen
**clear up** **71t** (auf)klären
**clearance** **70** Zwischenraum, Spiel, Abstand
**clip** **75** Klammer, Halterung, Rasthaken
**clip-on microphone** **76** Ansteckmikrophon
**closing date** **79** Einsendeschluss
**cloth** **24** Stoff, Tuch, Lappen
**clumsy** **54** ungeschickt
**CNC (Computerized Numerical Control)** **7** Computerisierte Numerische Steuerung
**coil** **60** Windung, Spule
**college of further education** **25** Berufsschule, Fachhochschule (für die Berufsausbildung)
**column** **24** Spalte, Kolumne
**come along** **59** mitkommen
**come loose** **37** sich lockern
**come up to** **37** herkommen
**come up with** **45** erfinden, sich ausdenken
**come with** **27** geliefert werden mit
**comfortable** **10** bequem, behaglich
**comment on** **67** kommentieren
**commercial** **19** Handels-, kommerziell
**commission** **46** übergeben, in Betrieb nehmen
**common** **11** normal, üblich
**communicate** **45** sich verständigen, kommunizieren
**communication** **25** Kommunikation
**communication processor card** **71t** Kommunikationsprozessorkarte
**communicator** **79** Kommunikator/in
**compact** **60** kompakt
**compare** **20** vergleichen
**compatibility** **67** Kompatibilität
**compatible** **77** kompatibel, zueinander passend
**complain** **11t** (sich) beschweren
**complaint** **11t** Beschwerde, Beanstandung, Reklamation
**complex** **43** kompliziert, komplex
**complicated** **64t** kompliziert
**component** **23** (Bau-)Element, (Bau-)Teil
**comprehensive school (GB)** **80** Gesamtschule
**compress** **63** komprimieren, zusammenpressen
**compressed** **24** Druck-
**compressed air** **51** Druckluft
**compressed-air pistol** **46** Druckluftpistole
**compression moulding** **46** Druckspritz-
**compression moulding unit** **46** Druckspritzmaschine
**computer-aided** **25** computergesteuert, -gestützt
**computerized** **42** Computer-, computerisiert
**conclude** **85p** (ab)schließen
**conclusion** **76** Ende, Abschluss
**condition** **39** Bedingung
**conduct** **44** leiten
**conductor** **32** Leiter
**conference** **45** Konferenz, Tagung
**confirm** **46** bestätigen
**conform to** **27** entsprechen
**conformity** **50** Übereinstimmung
**connect (to)** **28** verbinden (mit), anschließen (an)
**connected by** **60** verbunden durch
**connecting joint** **75** Verbindungsstück, Halterung
**connection** **49** Verbindung, Anschluss
**connector** **49** Verbindung(sstück)
**consider** **82** berücksichtigen, in Frage kommen
**take into consideration** **50** berücksichtigen
**consist of** **60** bestehen aus
**constant** **33** konstant
**constantly** **73** ständig
**construct** **49** bauen, konstruieren
**consulate** **87** Konsulat
**consultant** **65** Berater/in
**contact** **28** Kontakt
**contact detail** **81** Kontakt(adresse)
**contain** **15** enthalten
**contaminated** **86** verunreinigt, verschmutzt
**contents** **15** Inhalt
**continue** **65** fortfahren
**contractor** **46** Subunternehmer
**control valve** **62** Steuerventil
**control** **32** kontrollieren, regeln
**controller** **39** Steuer-, Bedienungsgerät
**convection** **70** Konvektion
**conveyor belt** **8** Fließ-, Förderband
**cooling** **51** Kühl-
**cooling system** **51** Kühlsystem
**corroded** **38** korrodiert, verrostet
**counter** **73** Zähler
**a couple of** **20t** ein paar
**coupling** **28p** Kopplung
**course** **25** Kurs, Lehrgang, Unterricht
**covalent bond** **41** Bindung durch gemeinsame Elektronen
**cover** **15, 75** (be-, ab)decken; Abdeckung
**covering letter** **79** Begleitbrief, Anschreiben
**CPU (Central Processing Unit)** **69** Zentraleinheit, -gerät, -baugruppe, Prozessor
**crane** **47** Kran
**create** **41** bilden, erzeugen
**critical** **73** kritisch, entscheidend
**CRT** **28p** CRT-Display, Flachbildschirm
**crystal** **41** Kristall
**cubed** **31** hoch drei
**current** **26** Strom
**currently** **13** momentan, zurzeit
**curve** **34** Kurve
**curve back/round** **34** sich zurückbiegen, krümmen
**customer relations** **59p** Kundenbeziehungen
**customer service** **69** Kundendienst

**cut out 65** ausschließen
**CV (curriculum vitae) 79** Lebenslauf
**cycle 60** Zyklus, Kreislauf
**cylinder 64t** Zylinder
**Czech Republic 16** Tschechische Republik
**damage 35** Schaden, Beschädigung
**data hold function 27** Funktion zur Datenspeicherung
**DC motor 60** Gleichstrommotor
**deadline 55** (Liefer-)Termin, Frist
**deal with 11t** sich befassen mit, bearbeiten, erledigen
**decide 16** (sich) entscheiden
**dedicated 25** engagiert
**defect 35** Defekt
**defective 35** fehlerhaft
**define 32** bestimmen, festlegen
**definitely 57** sicher(lich), auf jeden Fall
**definition 32** Definition
**delay 47** Verspätung, Verzögerung
**be delayed 10** Verspätung haben
**deliver 46** (aus)liefern
**delivery 46** (Aus-)Lieferung
**delivery bay 17** Lieferzone
**delta connection 61** Dreieckschaltung
**demand 35** Nachfrage, Bedarf
**demonstrate 79** zeigen, vorstellen, vorführen
**demonstration 20** Demonstration, Vorführung
**department 11t** Abteilung
**depending on 25** je nach(dem)
**depth 25** Tiefe
**design 13** konstruieren, entwerfen, entwickeln
**designated 14** bestimmt, gekennzeichnet
**designation 86** Bezeichnung
**designed for 70** bestimmt für
**desire 60** wünschen, verlangen
**despite 55** trotz
**detailed 76** detailliert, ausführlich, genau
**detect 35** entdecken
**determine 26** feststellen
**develop 19** entwickeln
**device 41** Vorrichtung, Gerät
**diac 36** Diac
**diagnose 34t** diagnostizieren
**dial 64t** (Wähl-)Scheibe
**difficulty 28** Schwierigkeit
**digit 28** (einstellige) Zahl, Ziffer
**digital multimeter 27** digitales Vielfachmessinstrument
**dimension 28p** Abmessung, Maß
**diode 27** Diode
**direction 19** Anweisung
**dirt 85** Schmutz
**disable 75** abschalten, deaktivieren
**discharging system 49** Entladungssystem
**discrete I/O point 69** digitaler Ein- und Ausgang
**discuss 23** besprechen, diskutieren (über)
**discussion 35** Diskussion, Gespräch
**disengage 70** ausschalten, ziehen
**disorganized 66** chaotisch
**dispatch 11t** Versand, Auslieferung
**dispatch supervisor 11** Versandleiter/in
**display 58** Bildschirm, Anzeige
**disposable 65** Einweg-, Wegwerf-
**dispose of 65** beseitigen, entsorgen
**dissipation 36** Ableitung
**distance 86** Entfernung, Abstand
**distract 14** ablenken
**distribute 41t** vertreiben, (aus)liefern
**divided by 31** geteilt durch
**doable 47** machbar
**doping 41** Doping, Verunreinigen
**dot 28** Punkt
**double back on itself 34** kehrtmachen, zurückkehren
**down there 11t** dort unten
**downtime 57** Ausfallzeit, Produktionsleerlauf(zeit)
**drastic(ally) 33** drastisch
**drawing 77** Zeichnung
**dress 15** verbinden
**drill 9** (durch)bohren
**driven by 58** angetrieben durch
**driving licence 79** Führerschein
**drop 33** fallen, sinken
**dry out 46** (aus)trocknen
**Dual Channel High sensitivity 28p** Hohe Zweikanalempfindlichkeit
**due to 55** wegen, aufgrund von
**duplicate 65** duplizieren
**duplication 65** Verdoppelung
**dust 15** Staub
**each other 28** einander, gegenseitig
**ear protection 14** Gehörschutz
**earn 78** verdienen
**easy enough 19t** ganz leicht
**effective(ly) 79** effektiv, wirksam, wirkungsvoll
**effectiveness 51** Effektivität, Wirksamkeit
**efficient(ly) 18** effizient, (leistungs)fähig
**electric 26** elektrisch
**electric current 60** (elektrischer) Strom
**electrical circuit 23** Stromkreis
**electrical discharge machine 8** Funkenerosionsmaschine
**electrical supply 51** Netz-, Stromanschluss
**electrician 25** Elektriker/in, Elektrotechniker/in
**electricity 15** Elektrizität, Strom
**electricity supply 49** Netz-, Stromanschluss
**electrics 37** Elektrik
**electrolytic capacitor 42** Elektrolytkondensator
**electromagnet 60** Elektromagnet
**electromechanical technician 78** Elektromechaniker/in
**electron 41** Elektron
**electronic 26** elektronisch
**electronic card 24** elektronische Karte
**electronics engineer 78** Elektroniker/in
**electronics 38** Elektronik
**electrostatic charge 49** elektrostatische Ladung
**elementary school (US) 80** Grundschule
**elevator 46** Aufzug
**eliminate 35** eliminieren, beseitigen
**else 11t** sonst (noch)
**embassy 87** Botschaft
**emergency 14** Notfall
**emergency stop 51** Notschalter
**emit 85** aussenden
**employ 25** ein-, anstellen, beschäftigen
**employee 7** Angestellte/r
**employer 25** Arbeitgeber/in
**employment 81** Beschäftigung, Anstellung

**enclose 81** beilegen, beifügen
**enclosed 82** Anlage
**endanger 51** gefährden
**ending 38** Endung, Ende (der Geschichte)
**energy 26** Energie
**engage 70** einstellen, -schalten, einstecken
**engineer 17** Ingenieur/in, Techniker/in
**engineering 6** Maschinenbau
**enormous 45** enorm, gewaltig
**ensure 35** sicherstellen, gewährleisten
**entertain 45** unterhalten
**entrance qualification 80** Reife
**environment 35** Umwelt, Umfeld
**envisage as 73** sich vorstellen als
**equal opportunities 79** Gleichberechtigung
**equally 37** gleichermaßen
**equals 31** ist gleich
**equation 31p** Gleichung
**equip 49** ausrüsten, ausstatten
**equipment 14** Maschinen, Geräte
**equivalent 80** Entsprechung, Übersetzung
**be equivalent to 25** entsprechen
**error 35** Fehler, Irrtum, Versehen
**especially 11t** besonders
**essentially 60** im Wesentlichen, im Grunde
**ethnicity 79** ethnische Zugehörigkeit
**EU (European Union) 81** Europäische Union
**even though 83** obwohl ja, selbst wenn
**event 20** Ereignis
**eventually 35** schließlich, zum Schluss
**everyday life 45** Alltag(sleben)
**examine 73** untersuchen, prüfen
**exceed 60** übersteigen, -treffen
**excellent 21** ausgezeichnet, hervorragend
**excerpt 48** Auszug
**exciter field 60** Erregerfeld
**execute 70** ausführen
**execution 73** Aus-, Durchführung
**exhibition 30** Ausstellung
**existing 75** bestehend, vorhanden
**expand 64** (sich) ausdehnen
**expansion 67** Erweiterung
**expansion connector 70** Erweiterungsverbindung
**expansion module 70** Erweiterungsmodul
**expect 41** erwarten, schätzen, annehmen
**experience 77** (Berufs-)Erfahrung
**experienced 29** erfahren
**explanation 52** Erklärung
**express 32** ausdrücken
**expression 15** (Gesichts-)Ausdruck
**ext. (extension) 28** Durchwahl, Apparat
**extensive 82** umfassend, ausgedehnt
**extreme 55** extrem
**have an eye for detail 79** Details wahrnehmen
**eye pad 15** Augenverband
**eye wash 15** Augenspülung
**facility 52** Vor-, Einrichtung
**factor 78** Faktor
**factory floor 37** Produktion(sfläche)
**fail to do 55** nicht tun
**failed 74** misslungen
**failing this 49** falls nicht
**failure 55** Versagen, Fehler
**be familiar with 7** vertraut sein mit, kennen
**FAQs (Frequently Asked Questions) 69** häufig gestellte Fragen
**fasten 15** befestigen, anlegen
**fast-growing 79** rasch wachsend
**fault 34** Fehler, Defekt
**fault-finding 37** Fehlersuche
**faulty 26** fehlerhaft, defekt
**feature 28p** Merkmal
**feedback 57** Rückmeldung
**feel comfortable with 45** sich gut fühlen mit
**fibre optic cable 46** Glasfaserkabel
**file 13** ablegen
**fill in 37** ausfüllen
**filler 9** Spachtel(masse)
**filter cartridge 63** Filterpatrone
**filter mat 24** Filter(einsatz)
**final 83t** Abschluss-
**finance 16** Finanz(en), Buchhaltung
**find out 26** herausfinden, feststellen
**finish off 20t** beenden
**fire extinguisher 14** Feuerlöscher
**first aid 14** erste Hilfe
**first of all 11t** zu(aller)erst, zunächst
**first thing 20t** als Allererstes, früh am Morgen
**fit 35** passen (zu), entsprechen
**fit in 11t** (sich) einfügen (in), gehören
**fit with 49** ausstatten mit
**fittings 63** Anschlüsse
**fix 36** reparieren, in Ordnung bringen
**fixed to 44** befestigt an
**flash 21** blinken, leuchten
**flashing mode 86** Blinkmodus
**flat-bladed screwdriver 22** Schlitzschraubenzieher
**flatten out 33** (sich) abflachen
**flexibility 69** Flexibilität
**flow chart 73** Flussdiagramm
**fluctuate 33** schwanken
**fluid 49** fließend
**fly on 50** weiterfliegen
**focus on 67** sich konzentrieren auf
**folding rule 22** Zollstock
**for short 11t** abgekürzt
**in force 49** in Kraft
**forecast 55** vorhersagen
**foreign 83t** ausländisch
**forklift 47** Gabelstapler
**formal 7** formell, förmlich
**formula (pl. formulae) 32** Formel
**forward voltage 34** Durchlassspannung
**forward-biased 34** im Durchlassbereich
**foundation 46** Fundament
**frequency 86** Frequenz
**frontliner 65** Vorreiter/in, Vorkämpfer/in
**full automation 17** volle Automation
**full/part-time 25** ganz-/halbtags
**function 19** Funktion
**function generator 26** Funktionsgenerator
**gain 73** gewinnen, erlangen
**gauze pad 15** Mullverband

**GCSE (General Certificate of Secondary Education) 80** mittlerer Schulabschluss

**gears 40** Getriebe

**gender 79** Geschlecht

**general 73** allgemein, generell

**generate 26** erzeugen

**germanium 41** Germanium

**gesture 45** Geste(n), Gestik

**get back to sb about sth 28** sich wieder bei jdm wegen etw melden

**get back to sb 37** jdn zurückrufen

**get on with sb 39** sich mit jdm verstehen

**get on with 11t** auskommen mit, sich verstehen mit

**get one's hands on sth 67** mit etw arbeiten

**get started 10** anfangen

**get sth done 52t** etw erledigen

**get sth off to a good start 83t** einen guten Start für etw finden

**get through to sb 52t** jdn erreichen

**get to know 6** kennen lernen

**get to 19t** gehen zu, kommen zu

**get used to 11t** sich gewöhnen an

**get wrong 66** missverstehen

**give off 36** ausstrahlen

**given 26** gegeben

**glad 20t** froh

**be go 47** funktionieren, arbeiten

**go about sth 86** in einer Sache vorgehen

**go across 19t** hinübergehen

**go for sth 27** etw nehmen, sich für etw entscheiden

**go off to 44** abgehen zu

**go on with 73** fortfahren mit

**go over 11t** durchgehen

**go wrong 52** schiefgehen

**goggles 14** Schutzbrille

**be good at sth 68** gut in etw sein, etw gut können

**gradual(ly) 33** allmählich

**grammar school (GB) 80** Gymnasium

**graph 33** Kurve, grafische Darstellung, Grafik

**graphical 77** grafisch

**grasp 83t** erfassen, begreifen, verstehen

**grease 14** Schmiere, Fett

**greater/larger than 31** größer als

**greetings 59** Grüße

**grip 23** festhalten, (er)greifen

**guarantee 35** Garantie

**guide 39** führen, lenken

**guideline 14** Richtlinie, Hinweis

**hacksaw 22** Bügelsäge

**hammer 22** Hammer

**hand movement 66** Handbewegung

**handbook 57** Bedienungsanleitung, Handbuch

**handheld 39** Hand-(Gerät), tragbar

**handle 13** erledigen, bearbeiten

**handling 51** Handhabung, Umgang

**hands-on 67** aktiv, (selbst) zupackend

**handy 27** zur Hand, parat

**hang off 37** (lose) hängen an

**hang on 27** dranbleiben

**happen 20** passieren, geschehen

**hard hat 14** Schutzhelm

**hardwired 77** verdrahtet

**hardworking 82** fleißig

**have a go 64t** es versuchen

**hazard 49** Risiko, Gefahr

**hazardous 14** risikoreich, gefährlich

**head of accounts 11** Leiter/in Buchhaltung

**head of department 11t** Abteilungsleiter/in

**headquarters 7** Hauptsitz, Zentrale

**health and safety 14** Arbeitsschutz

**health insurance 87** Krankenversicherung

**heat exchanger 64** Wärmetauscher

**hectic 11t** hektisch

**helper 81** Helfer/in, Hilfskraft

**hesitate 82** zögern

**highly 82** äußerst, hoch

**high school (US) 80** weiterführende Schule, Haupt-, Realschule

**High School Diploma (US) 80** (entspricht) Abitur

**high-speed 60** Hochgeschwindigkeits-

**hit 33** erreichen

**HNC (Higher National Certificate) 25** staatliches Abschlusszeugnis

**hold 46, 62** (fest)halten; speichern, aufbewahren

**hold on 59** dranbleiben

**hold the line 28** dranbleiben

**hold up 52** aufhalten

**hole 23** Loch

**holster 27** Halfter, Tragetasche

**honoured 59** geehrt

**hook onto 75** einhängen, -haken

**horizontal(ly) 28p** horizontal, waagerecht

**hose 24** Schlauch

**housing 49** Verkleidung, Gehäuse

**How do you do? 7** Guten Tag/Morgen/Abend!

**however 20** doch, jedoch

**HR (Human Resources) Co-ordinator 83** Koordinator/in in der Personalabteilung

**HR (Human Resources) Manager 6** Personalleiter/in, -chef/in

**hugely 65** riesig, ungeheuer

**human 35** menschlich

**humid 85** feucht

**be in a hurry 19t** es eilig haben

**hydraulic 51** hydraulisch

**hydraulics 25** Hydraulik

**hyphen 28** Bindestrich

**I/O expansion cable 69** Erweiterungsskabel für Ein- und Ausgänge

**identify 22** identifizieren, feststellen

**illustrate 74** illustrieren, veranschaulichen

**imagine 11t** sich vorstellen

**immigration service 87** Einwanderungsstelle

**impact proof 86** stoß-, schlagfest

**importance 78** Wichtigkeit, Bedeutung

**impressed 84** beeindruckt

**impression 40** Eindruck

**impressive 17** beeindruckend, eindrucksvoll

**improve 35** verbessern

**improvement 35** Verbesserung

**impurity 41** Verunreinigung, Unreinheit

**in order to 45** um zu

**in the end 10** zum Schluss, schließlich

**in the event of 15** im Falle eines

**in the meantime** **83t** in der Zwischenzeit
**in time** **50** rechtzeitig
**inadvertent** **35** unbeabsichtigt, versehentlich
**inch** **28p** Inch, Zoll
**include** **43** angeben
**including** **13** einschließlich, inbegriffen, darunter
**incorrect** **35** falsch, unrichtig
**increase** **65** zunehmen, anwachsen, steigen
**increasingly** **25** zunehmend
**index card** **35** Karteikarte
**indicator** **64t** Anzeige
**induction AC motor** **60** Induktionswechselstrommotor
**induction motor** **60** Induktionsmotor
**induction programme** **7** Arbeitseinführungsprogramm
**inductive** **85** induktiv
**informal(ly)** **66** ungezwungen, nicht förmlich
**infra red** **86** infrarot
**on your own initiative** **79** selbständig
**injure** **14** verletzen
**injury** **14** Verletzung
**inlet** **51** Einlass, Zufluss
**inline skating** **81** Inlineskaten
**input** **28p** Eingang, Aufnahme
**input impedance** **28p** Eingangsimpedanz
**input/output control** **74** Eingangs- und Ausgangsregler
**insert** **75** einsetzen, einführen
**insight** **87** Einblick
**insist** **55** (darauf) bestehen
**inspection** **57** Kontrolle, Inspektion
**install** **21** installieren
**installation** **46** Installation, Auf-, Einbau, Montage
**installation configuration** **69** Konfiguration, Montageanordnung
**instruct** **14** anweisen, beauftragen
**instruction** **15** Anweisung, Hinweis
**instruction manual** **27** Handbuch
**insulate** **44** isolieren, abschirmen
**insulation** **54** Isolierung, Isolation
**insurance** **13** Versicherung
**intelligence** **35** Intelligenz
**intelligent** **35** intelligent
**interact with** **60** sich gegenseitig beeinflussen
**interaction** **60** Interaktion
**interchange** **77** austauschen
**internal graticule** **28p** internes Netz, Gitter, Koordinatensystem
**interrupt** **40** unterbrechen
**interruption** **41t** Unterbrechung
**interval** **24** Abstand
**interview** **83** Vorstellungsgespräch
**interviewee** **83** Gesprächs-, Interviewpartner/in
**interviewer** **85p** Leiter/in des Vorstellungsgesprächs, Interviewer/in
**intricate** **77** kompliziert
**introduce (oneself)** **7** (sich) vorstellen
**introduction** **6** Einführung
**invaluable** **87** unschätzbar, (sehr)wertvoll
**invent** **77** erfinden
**invest** **65** investieren, anlegen
**investment** **65** Investition
**involve** **36** beteiligen
**involved in** **48** beteiligt an, zu tun haben mit
**irritant** **15** Reizstoff
**irritate** **15** reizen
**in italics** **21** kursiv
**item** **15** Artikel, Gegenstand
**jerk** **51** Erschütterung, Ruck, Stoß
**jewellery** **14** Schmuck(stücke)
**jobsite** **87** Internetseite mit Stellenanzeigen
**join** **7** sich anschließen, nachkommen
**join up** **77** sich anschließen (lassen)
**joined to** **44** verbunden mit
**joint** **9** Gelenk
**junction** **43** Kreuzung, Verbindungsstelle, Knoten
**be keen to** **35** erpicht sein auf, sehr interessiert sein an
**keep clear** **55** freihalten
**keep in place** **15** befestigen
**keep up** **62** aufrecht erhalten
**key focus** **83t** Hauptschwerpunkt
**keyword** **77** Schlüssel-, Stichwort
**label** **43** beschriften, kennzeichnen
**ladder logic** **77** Strickleiterlogik
**language skills** **81** Sprachkenntnisse
**laser technician** **78** Lasertechniker/in
**laser welder** **8** Laserschweißgerät
**last for** **25** dauern
**latest** **17** neueste
**lathe** **7** Drehbank
**lattice** **41** Gitter
**Latvia** **79** Lettland
**law** **81** Gesetz
**layer** **36** Schicht
**layout** **21** Grundriss, Plan
**lead the way** **35** anführen, vorangehen
**leakage** **24** Leck(stelle)
**lean manufacturing** **35** schlanke Produktion
**learner** **82** Lerner/in
**leave sth to sb** **64t** jdm etw überlassen
**lecturer** **25** Dozent/in
**LED** **23** LED-Anzeige
**left-hand side** **44** linke Seite
**length** **51** Länge
**let sb know** **28** jdm Bescheid geben/sagen
**letter of application** **83t** Bewerbungsschreiben
**level** **11t** Ebene, Stufe
**level off** **33** sich einpendeln
**liaise with** **79** sich in Verbindung setzen mit
**lie behind** **65** stecken hinter
**life-saving** **15** Lebensrettungs-
**lifespan** **65** Lebenszeit, -dauer
**lift** **9** (an-, hoch)heben
**lifting device** **51** Hebevorrichtung
**light barrier** **86** Lichtschranke
**light beam** **85** Lichtstrahl
**light up** **45** aufleuchten
**like** **60** gleich
**limit** **36** begrenzen, beschränken
**limit switch** **85** Endlagenschalter
**line sth up** **47** etw organisieren
**liquid** **14** Flüssigkeit
**load** **9** Last
**load onto** **17** aufladen
**loading bay** **17** Ladebucht, -zone
**located** **64** gelegen

**location** 19 (Stand-)Ort, Lage, Position
**long term** 35 langfristig, Langzeit-
**look after** 8 sich kümmern um, betreuen
**look around** 35 (sich) umsehen
**look forward to** 10 sich freuen auf
**look over** 62 sich ansehen
**loose** 14 weit (geschnitten), lose
**loosen** 23 lockern, lösen
**lorry** 17 Lastwagen, Lkw
**lost** 19t verloren
**lower** 47 senken, herablassen
**Ltd (Limited)** 46 GmbH
**lubrication** 63 Schmierung
**machine tool** 8 Werkzeugmaschine
**machinery** 8 Maschinen(park), Geräte
**made from** 36 hergestellt aus
**be made up of** 36 zusammengesetzt sein aus, bestehen aus
**magnet** 60 Magnet
**magnetic** 60 magnetisch
**magnetic field** 60 Magnetfeld
**mainly** 36 hauptsächlich, vor allem
**the mains current** 60 Netzstrom
**mains frequency** 61 Netzfrequenz
**mains tester** 36 Strommesser
**maintain** 32 einhalten
**maintenance** 14 Wartung, Pflege
**maintenance check** 20 Wartung(stest)
**major** 15 groß, größer
**major artery** 15 Hauptarterie
**I can't make it** 57 es passt mir nicht
**make eye contact** 45 Blickkontakt herstellen
**make notes** 29 (sich) Notizen machen
**make sense** 38 sinnvoll sein
**make sth up** 53p etw erfinden, sich etw ausdenken
**make sure** 20t sicherstellen, sich vergewissern
**make up (the) time** 55 die Zeit einholen
**make up one's mind** 83t sich entschließen
**malfunction** 35 Fehlfunktion
**manage** 52t (es) schaffen
**mandrel** 46 Dorn, Spindel
**manual** 27 Handbuch
**manually-controlled** 9 von Hand bedient
**manufacture** 25 herstellen, produzieren, verarbeiten
**manufacturer** 65 Hersteller/in, Produzent/in
**manufacturing equipment** 79 Produktionsmaschinen
**marital status** 81 Familienstand
**mark** 15 kennzeichnen
**marker pen** 76 Folien-/Tafelstift
**marketing** 11t Marketing, Werbung, Vertrieb
**marketing manager** 11 Marketingleiter/in
**mathematical** 31 mathematisch
**matter** 65 Sache, Angelegenheit
**meaning** 14 Bedeutung
**means of transport** 51 Beförderungsmittel
**measure** 23 messen
**mechanic** 25 Mechaniker/in
**mechanical** 38 mechanisch
**mechanics** 37 Mechanik, Ablauf
**mechatronics** 6 Mechatronik
**mechatronics technician** 6 Mechatroniker/in
**mediation** 15 Vermittlung
**medical** 15 medizinisch, ärztlich
**melt** 9 schmelzen
**member** 71 Mitglied
**member of staff** 20 Mitarbeiter/in
**membership** 78 Mitgliedschaft
**memory** 39 Speicher
**mention** 14 erwähnen
**take a message** 28 etw ausrichten, eine Nachricht entgegennehmen
**microprocessor** 42 Mikroprozessor
**milling machine** 8 Fräsautomat
**mind** 35 hier: Meinung
**not mind** 41t nichts dagegen haben
**minor** 15 gering(fügig)
**minus** 31 minus
**have a minute** 27 eine Minute (Zeit) haben
**mirror** 34 (wider)spiegeln
**mirror image** 34 Spiegelbild
**misaligned** 86 Dejustage, falsch justiert
**missing** 38 fehlend, nicht vorhanden
**mix up** 51 verwechseln
**mobile phone** 52t Mobiltelefon, Handy
**mode** 28p Art, Modus
**model** 30 Modell, Gerät
**module** 67 Modul
**moist** 15 feucht, nass
**monitor** 69 überwachen, kontrollieren
**monitoring** 83 Überwachung, Kontrolle
**moreover** 82 außerdem, zudem
**motion** 39 Bewegung
**motivated** 79 motiviert
**motorway** 52t Autobahn
**mount** 51 montieren, einbauen
**mounting** 75 Montage-, Einbau-
**mounting configuration** 69 Einbaukonfiguration, Montageanordnung
**mouth-to-mouth resuscitation** 15 Mund-zu-Mund-Beatmung
**move on** 68 fortfahren
**movement** 39 Bewegung
**moving** 9 laufend
**moving part** 37 bewegliches Teil
**MRO (Maintenance, Repair and Overhaul)** 65 Wartung, Reparatur und Überholung
**multimeter** 22 Vielfachmessinstrument
**multiplied by** 31 multipliziert mit
**multipoint interface cable** 71t Mehrfachschnittstellenkabel
**mushroom valve** 64 Tellerventil
**my pleasure** 57 gern (geschehen)
**narrow** 27 eng
**nasty** 54 schlimm, unangenehm
**nationality** 81 Staatsangehörigkeit
**neat(ly)** 43 ordentlich, sauber
**necessary** 14 nötig, erforderlich
**needle nose pliers** 22 Spitzzange
**neighbouring** 41 angrenzend, benachbart
**newly qualified** 78 frisch qualifiziert, ausgebildet
**newly-designed** 26 neu konstruiert
**next door to** 17 neben
**nipple** 63 Nippel
**nitrogen** 64 Stickstoff
**no-load running** 86 Stromaufnahme im Leerlauf
**noon** 10 Mittag
**not in any way** 37 in keiner Weise

**notation 77** Aufzeichnung, Bezeichnung
**note 31** beachten
**notes 83** Notizen
**a number of 55** eine Reihe von
**nut 23** Mutter
**observation 77** Beobachtung
**observe 51** beachten
**obvious 65** offensichtlich
**occupation 25** Beruf, Tätigkeit
**occur 35** vorkommen, passieren, eintreten
**oh dear 53** ach du liebe Zeit!
**Ohm's law 32** das Ohmsche Gesetz
**ohmmeter 36** Ohmmeter
**OHP (overhead projector) 76** Overheadprojektor
**oil 14** Öl
**oil level indicator 64** Ölstandsanzeige
**ointment 15** Salbe
**omit 43** aus-, weglassen
**on board 47** an Bord
**on hand 50** vorhanden
**on the other side of 19** gegenüber
**on time 55** pünktlich, rechtzeitig
**on top of 78** zusätzlich
**on top of that 11t** darüber hinaus
**once again 76** noch einmal
**one by one 41t** einer nach dem anderen
**operate 14** bedienen, betätigen
**operating 51** Arbeits-
**operating cycle 73** Ablaufprogramm
**operating manual 63** Handbuch, Betriebsanleitung
**operating panel 46** Betriebs-, Schalttafel
**operation 31** Operation
**operational 47** einsatzfähig, in Betrieb
**opportunity 78** Chance, Möglichkeit
**opposing 60** gegenüber (liegend)
**optional 69** wahlweise, optional
**optoelectrical 85** optoelektrisch
**orbital 41** Umlaufbahn
**order 10** Reihenfolge
**organization chart 12** Organigramm
**organize 13** organisieren
**orient 75** ausrichten
**originally 46** ursprünglich
**oscilloscope 23** Oszilloskop
**outer 41** äußere, Außen-
**outline 67** umreißen, skizzieren
**output 28p** Ausgang, -gabe
**over and over again 39** immer wieder
**over there 6** dort (drüben)
**over to sb 11** jd ist dran
**overall 12** (ins)gesamt, Gesamt-, umfassend
**overdue 52t** überfällig
**overhaul 65** Überholung
**overhauled 65** überholt
**overland 50** über Land
**overload protection 27** Überlastungsschutz
**overnight 30** über Nacht
**oversee 53p** beaufsichtigen
**overview 41t** Überblick, -sicht
**overvoltage 27** Überlastung
**owing to 49** wegen, aufgrund
**package 17** verpacken
**packaging 11t** Verpackung
**packaging robot 37** Verpackungsroboter
**pad 51** Polster, Schützer
**a pair of pliers 23** eine Zange
**in pairs 28** paarweise
**panel 69** Platte, Schalttafel
**paragraph 45** Absatz, Abschnitt
**in parallel 36** parallel geschaltet
**on sb's part 55** auf jds Seite
**particular 58** bestimmt
**particularly 40** besonders
**pass 63** leiten
**pass along 65** transportiert werden
**pass through 62** hindurchleiten
**passage 14** Gang, Durchgang
**pastime 81** Freizeitaktivität
**pay 11t** Lohn, Gehalt, Entlohnung, Bezahlung
**payment 13** Zahlung, Auszahlung
**peak 33** gipfeln, den Höchstwert erreichen
**pension cover 87** Rentenanspruch, -zahlung
**perfectionism 83** Perfektionismus
**perfectionist 83** Perfektionist/in
**perform 19** durchführen
**performance 61** Leistung
**period 58** (Zeit-)Raum
**permanent 60** ständig, permanent
**persist 15** andauern
**personal 81** persönlich, privat
**personal detail 81** persönliche Angabe
**personal injury 75** Körperverletzung, Personenschaden
**personal quality 79** persönliche Eigenschaft
**personal space 45** Diskretionsabstand
**personnel 11t** Personal, Mitarbeiter
**phase error 28p** Phasenverschiebung
**Phillips screwdriver 22** Kreuzschlitzschraubendreher
**philosophy 35** Philosophie
**phrase 12** (Rede-)Wendung, Ausdruck
**pick and place robot 45** Bestückungs-/Montageroboter
**pick 9** auswählen, mitnehmen
**pick up 23** aufheben
**get a picture of 41t** sich ein Bild machen
**pie chart 32** Kreisdiagramm
**pin 36** heften
**pipe 49** Rohr, Röhre
**piston 85** Kolben
**placement 81** Praktikum
**plant 65** Maschinen, Anlage
**play a role in 65** eine Rolle spielen bei
**PLC (Programmable Logic Controller) 25** speicherprogrammierbare Steuerung
**plc (public limited company) 20p** AG
**PLC unit 72** PLC-Einheit
**plunge 33** (ab)stürzen, absacken
**plus 31** plus
**pneumatic 51** pneumatisch
**pneumatic cylinder 85** Druckzylinder
**pneumatics 25** Pneumatik
**pneumatics line 51** Druckluftleitung
**point 19** Punkt, Stelle
**polarity 60** Polarität
**pole 60** Pol
**policy 78** (Firmen-)Politik

**polite 11t** höflich
**pollution 86** Verschmutzung
**popular 65** beliebt, populär
**port 70** Anschluss, Buchse, Schnittstelle
**Pos. (position) 63** Stelle, Position
**postpone 55** verschieben
**potential 32** Potenzial
**potential(ly) 49** potenziell
**power cord 28p** Netz-, Stromkabel
**power drill 22** elektrische Bohrmaschine, Elektrobohrer
**power factor 61** Leistungsfaktor
**power source 28p** Netzanschluss
**power supply 43** Netz-, Stromanschluss, Spannungsversorgung
**powered 23** elektrisch (angetrieben)
**practise 11** (ein)üben, trainieren
**prefer 68** vorziehen, lieber mögen, bevorzugen
**prepare 66** vorbereiten
**presence 26** Anwesenheit
**(give a) presentation 66** Vorstellung, Präsentation (eine ~ durchführen)
**presenter 71** Moderator/in
**press 51** drücken
**pressure 15** Druck
**pretty 17** hier: ziemlich
**prevent 35** (ver)hindern, verhüten
**preventable 48** vermeidbar
**prevention 51** Verhütung
**previous 73** vorhergehend, vorherig
**primary school (GB) 80** Grundschule
**primary production 17** Grundstoffproduktion
**principle 60** Grundsatz, Prinzip
**probe 27** Sonde
**problem-solving 67** Problemlösung
**procedure 15** Vorgehensweise, Maßnahme
**proceed 75** (fort)fahren, vorgehen
**process 17** (weiter)verarbeiten, bearbeiten
**process of elimination 86** Ausschlussverfahren
**produced by 60** erzeugt durch
**production engineer 16** Betriebsingenieur/in
**production hall 17** Produktionshalle
**production line robot 7** Fertigungsbandroboter
**production line 11t** Fertigung(sband), Produktion
**production manager 11** Produktionsleiter/in
**production smoothing 35** Produktionsglättung
**production supervisor 7** Produktionsleiter/in
**profitable 65** gewinnbringend, rentabel
**programme into 35** einprogrammieren
**programmer 39** Programmierer/in
**programming 67** Programmieren
**project management 25** Projektmanagement
**promise 28** versprechen, zusagen
**promotional prospect 78** Berufsaussichten, -chancen
**pronounce 31** aussprechen
**proper(ly) 14** richtig, korrekt, ordnungsgemäß
**protect 15** schützen
**protective 86** Umgebungs-
**proud 35** stolz
**provide 48** zur Verfügung stellen, erbringen
**proximity 85** Nähe
**pry sth away 75** etw aufhebeln
**pulley 8** Flaschenzug, Rolle
**pulse 15** Puls
**pulsed 86** gepulst
**pump 56** pumpen
**pump pliers 22** Wasserpumpenzange
**pure(ly) 45** rein
**put sb through 26** jdn durchstellen, jdn verbinden
**put together 65** zusammensetzen
**quality control 11t** Qualitätskontrolle
**quartz 41** Quarz
**in question 87** in Frage kommend, betreffend
**R&D (research and development) 17** Forschung und Entwicklung
**racial discrimination 81** Rassendiskriminierung
**rail 69** Schiene
**range 27** Bandbreite, Bereich
**rank 78** einstufen, anordnen
**rate 26** Rate
**rated speed 61** Nenndrehzahl
**rather (than) 20** anstatt
**rating plate 61** Typenschild
**raw material 17** Rohstoff(e)
**read back 28** (noch einmal) vorlesen
**read out 31p** (laut) vorlesen
**realistic(ally) 87** realistisch(erweise)
**in reality 73** in Wirklichkeit, tatsächlich
**realize 65** verwirklichen, umsetzen
**rear 55** Rückseite
**rearrange 72** (um)ändern
**reasonable 35** angemessen, günstig
**reassure 52** versichern
**receiver 85** Empfänger
**recently 83** kürzlich, vor kurzem
**receptionist 6** Empfang(smitarbeiter/in)
**recertify 65** (im Werk) komplett überholen, neu aufbauen
**recognize 25** (an)erkennen
**record 73** aufnehmen, -zeichnen
**record card 35** Karteikarte
**recording 22** Aufnahme, Aufzeichnung
**recovery 15** Erholung
**recruit 13** einstellen
**rectangular 28p** rechteckig
**rectify 52** richtigstellen, korrigieren
**refer to 64** sich beziehen auf
**referee 79** Referenz(geber)
**reference 81** Referenz, Empfehlungsschreiben
**refit 48** wieder einsetzen
**reflect 25** (sich) widerspiegeln
**reflection angle 86** Abstrahlwinkel
**regardless of 79** ungeachtet, ohne Rücksicht auf
**in the region of 79** im Bereich von
**regulation 51** Bestimmung, Verordnung, Vorschrift
**regulator 62** Regler
**relate to 45** eine Beziehung haben zu, reagieren auf
**related 75** (miteinander) verbunden, angeschlossen
**relating to 48** in Bezug auf, in Zusammenhang mit, betreffend
**in relation to 74** in Bezug auf, im Verhältnis zu
**relationship 12** Beziehung, Verhältnis
**relay 77** Relais
**release 62** Zufuhr
**relief valve 51** Entlastungsventil
**remain 33** bleiben

**remaining 32** restlich, verbleibend
**remanufacture 65** (vor Ort) warten und reparieren
**remind 50** (daran) erinnern
**remove 15** entfernen, beseitigen
**render 49** machen
**repair 6** reparieren
**repair brief 37** Reparaturanleitung, Serviceunterlagen
**repair job 59p** Reparaturarbeit
**repeat 39** wiederholen
**repel 60** (sich) abstoßen
**replace 30** ersetzen
**replacement 52** Ersatz(teil)
**replacement part 53p** Ersatzteil
**report 13** berichten
**report to 12** unterstehen, unterstellt sein
**representative 41t** Vertreter/in, Repräsentant/in
**request 40** Bitte, Anfrage, Wunsch, Anweisung
**require 61** benötigen
**required 77** verlangt, erforderlich
**requirement 65** Bedingung, Anforderung
**reschedule 54** neu planen, ansetzen
**research 87** Nachforschung(en)
**resemble 77** ähneln, gleichen
**resident 45** wohnhaft, ansässig
**residual ripple 86** Restwelligkeit
**resistance 26** Widerstand
**resistor 23** Widerstand
**re-solder 37** anlöten
**resources 12** Mittel, Ressourcen
**respect 39** respektieren
**respective 76** jeweilig
**response 10** Reaktion, Antwort
**in response to 35** als Reaktion auf, entsprechend
**responsibility 78** Verantwortlichkeit, Verantwortung (für)
**take responsibility for 55** (die) Verantwortung übernehmen für
**responsible for 7** verantwortlich/zuständig für
**responsible to sb 11t** jdm unterstellt
**result in 31** ergeben
**retraining 65** Umschulung
**get a return on sth 65** einen Gewinn durch etw erzielen
**return flight 59p** Rückflug
**return line 51** Rückleitung
**reversal 60** Umkehrung
**reverse 60** umkehren
**reverse voltage 34** Sperrspannung
**in reverse order 81** in umgekehrter Reihenfolge
**reversed-biased 34** in Sperrrichtung
**review 76** wiederholen, zusammenfassen
**revolution 60** (Um-)Drehung
**rework 77** überarbeiten
**rhetoric 81** Rhetorik
**right away 21** sofort, umgehend
**right now 19t** sofort
**rise time 28p** Anstiegszeit
**robotic 39** Roboter-
**roboticist 45** Roboteringenieur/in
**robotics 17** Robotik, Automatisierung
**robotics engineer 78** Roboteringenieur/in
**role-play 28** als Rollenspiel durchführen
**rope 9** Seil, Tau
**rotate 60** rotieren, (sich) drehen
**rotating 86** drehbar
**rotation 51** Drehung
**rotor 60** Rotor, Läufer
**roughly 25** ungefähr, etwa
**round nose pliers 22** Rundzange
**route 25** Route, Weg
**routine 20** Routine(arbeit), routinemäßig
**row 17** Reihe
**rpm (revolutions per minute) 61** UpM (Umdrehungen pro Minute)
**rule 15** Vorschrift, Regel
**rung 77** Sprosse
**safety boot 14** Sicherheitsschuh
**safety device 51** Sicherheitsvorrichtung
**safety glove 14** Schutzhandschuh
**safety guideline 48** Sicherheitsbestimmung
**safety measure 51** Sicherheitsmaßnahme
**safety pin 15** Sicherheitsnadel
**safety regulation 52** Sicherheitsbestimmung
**salary 13** Gehalt, Lohn
**sales 11t** Vertrieb, Verkauf
**sales department 13** Verkauf(sabteilung)
**saving 65** Einsparung, Ersparnis
**scan 36** überfliegen, durchsehen
**scanning equipment 79** Scanner(geräte)
**schedule 47** (Zeit-)Plan
**on schedule 53p** nach Plan, pünktlich
**scheduled 55** vorgesehen, geplant, planmäßig
**schematic 77** schematisch(e Darstellung)
**scheme 25** Plan, Programm
**scientific 60** wissenschaftlich
**scrap 65** verschrotten, entsorgen
**scrape 38** kratzen, schaben, reiben
**screen 27** Bildschirm
**screw 23** Schraube
**screwdriver 22** Schraubenzieher, -dreher
**seamless(ly) 77** nahtlos
**secondary 42** sekundär, Neben-
**secondary school (GB) 80** weiterführende Schule, Haupt-, Realschule
**secondary production 17** verarbeitende Produktion
**section 11t** Abteilung
**secure 51** absichern, festmachen
**secure(ly) 75** sicher, fest
**seek 15** (auf)suchen
**seeming(ly) 35** scheinbar, anscheinend
**segment 32** Teilstück, Segment
**select 79** auswählen
**selection 27** Auswahl
**self-explanatory 65** unmittelbar verständlich
**semi-conducting 36** Halbleiter-
**semiconductor 41** Halbleiter
**seminar 59p** Seminar, Fachtagung
**send out 37** verschicken, aussenden
**senior 20** vorgesetzt, leitend, ranghöher, älter
**sensitivity 28p** Empfindlichkeit
**sensitivity setter 86** Empfindlichkeitseinsteller
**sensor 25** Sensor, Messgerät, Melder, (Mess-)Fühler
**sensor/actuator box 64** Sensor-/Aktor-Box
**sensor application 79** Sensoreinsatz
**sensor head 85** Sensorkopf
**sensor system 79** Sensorsystem

**sensor technician 78** Messtechniker/in
**separately 28** getrennt
**sequence 19** Reihe(nfolge)
**sequential 77** sequenziell, aufeinander folgend
**a series of 77** Serie, Reihe
**in series 36** in Reihe geschaltet
**serious(ly) 14** ernst(haft)
**servant 45** Diener/in
**service 21** Service, Wartung, Inspektion
**session 37** Sitzung
**set 51** einsetzen, -stellen
**a set of 77** eine Reihe von
**set up 45** veranstalten
**severe 15** schwer, stark
**sexual orientation 79** sexuelle Orientierung
**shaft 60** Achse, (Haupt-)Welle
**shake hands 11** sich die Hand geben
**shape 9** formen
**share responsibility for 55** Mitverantwortung übernehmen für
**sharp(ly) 33** rasch, schnell, stark
**sheet 28** Blatt
**shift 24** Schicht
**shock 15** Schock, Schlag
**shoot up 33** in die Höhe schießen
**shopfloor 7** Produktion(sbereich)
**short-circuit 37** einen Kurzschluss auslösen
**shortcut 14** Abkürzung
**show sb around 20t** jdn herumführen
**shut down 14** aus-, abschalten
**side cutting pliers 22** Seitenschneider
**sign 14** Schild
**signal for 35** herbeirufen
**signposting phrase 76** strukturierender Ausdruck
**silicon 36** Silizium
**be similar to 25** ähneln
**similarity 77** Ähnlichkeit
**single 81** alleinstehend, unverheiratet
**site 16** (Arbeits-)Stelle, (Firmen-)Gelände, Anlage
**on site 65** vor Ort
**site plan 18** Lageplan
**skill 25** Fertigkeit, Fähigkeit
**skilled 51** ausgebildet, Fach-
**skin 15** Haut
**slide 66** Dia(positiv), Folie
**slot-head screwdriver 22** Schlitzschraubenzieher
**slow down 57** langsamer werden
**small talk 10** höfliche Konversation, Smalltalk
**smaller than 31** kleiner als
**smoulder 15** schwelen, glimmen
**snap into place 75** (richtig) einrasten
**snap sth closed 75** etw schließen, einrasten lassen
**snap sth open 75** etw öffnen
**solder 6** löten
**soldering iron 22** Lötkolben
**solid 60** fest
**solve 36** lösen
**some months back 37** vor einigen Monaten
**sometime 27** irgendwann
**sophisticated 77** hoch entwickelt, differenziert, kompliziert
**sort 45** Art, Sorte
**sort into 70** sortieren nach
**sort out 11t** in Ordnung bringen, klären, lösen
**sound 16** (sich) anhören, klingen
**sound as if 57** klingen als ob
**source 15** Quelle
**space 70** Raum, Platz
**spanner 22** Schraubenschlüssel
**spare 27** Ersatz-
**specialize in 79** sich spezialisieren auf
**specific 49** speziell
**specification 20t** Angabe, Aufstellung
**specification sheet 86** Spezifikationsblatt
**specifications (specs) 27, 37** genaue Angaben, Spezifikationen
**speech 45** Sprache
**speed 60** Geschwindigkeit
**spill 14** ausschütten, verschütten
**spin 9** (schnell) drehen
**splendid 57** prima, ausgezeichnet
**split-ring commutator 60** Stromwender mit Stromwenderlamellen
**spoken 72** gesprochen, mündlich
**spontaneous 83t** spontan
**spot on 71t** haargenau richtig
**square root of 31** Quadratwurzel
**squared 31** hoch zwei
**squarewave 28p** Rechteckwelle
**squeeze 65** herausholen
**staff 8** Angestellte, Personal
**stage 73** Phase, Abschnitt, Stadium
**stand one's ground against 35** seine Position behaupten gegen(über)
**stand-alone 69** allein stehend
**standard rail 75** Hutschiene
**star connection 61** Sternschaltung
**start (sth) up 21, 48** starten, in Betrieb setzen
**starting point 65** Ausgangspunkt
**state 57** feststellen, angeben
**statement 8** Aussage, Feststellung
**static 49** statisch
**stationary 60** fest(stehend), Stand-
**stator 60** Ständer, feststehender Teil
**status 70** Status, Stand
**status display 86** Statusanzeige
**steady (-ily) 33** gleich-, regelmäßig, stetig, ständig
**step by step 87** Schritt für Schritt
**sterile 15** steril
**STL instruction 74** STL-Anleitung
**storage 11t** Lager(haltung)
**store 19** lagern, aufbewahren
**storm 11t** stürmen
**straight 43** gerade
**straight away 15** sofort, gleich, auf der Stelle
**straight on 19t** geradeaus
**straightforward 64t** einfach
**strategy 65** Strategie
**strength 25** Stärke
**strict 12** streng
**strike 53p** Streik
**strip 23** abisolieren
**structure 11** Struktur, Aufbau
**stuck 37** fest-, eingeklemmt
**study for 25** studieren

**subsidiary 6** Tochter(gesellschaft)
**substance 19** Substanz, Stoff
**successful 50** erfolgreich
**such as 25** wie (zum Beispiel)
**suck 46** saugen
**suddenly 33** plötzlich
**suggest 35** vorschlagen
**suggestion 53** Vorschlag
**suit sb 57** jdm passen, jdm recht sein
**suitable 30** geeignet, passend
**suited to 80** geeignet für
**summarize 37** zusammenfassen
**summary 76** Zusammenfassung
**summer camp 81** Sommer(ferien)lager
**supervisor 12** Leiter/in
**supplier 53p** Lieferant/in, Zulieferer
**supplies 19** Vorräte, Material
**supply 15** versorgen
**supply current 60** Stromzuleitung
**supply with 60** (be)liefern mit
**support 79** verstärken, unterstützen
**suppose 11t** glauben, annehmen
**surface 32** (Ober-)Fläche
**surplus 24** überschüssig
**surprising(ly) 83** überraschend(erweise)
**swap 25** (ver)tauschen
**sweep time 28p** Abtastzeit
**swing 34** schwingen
**switch 36** Schalter
**switch on 30** an-, einschalten
**switching unit 85** (Um-)Schaltgerät
**synchronization 28p** Synchronisierung
**synchronous AC motor 60** Synchronwechselstrommotor
**synchronous motor 60** Synchronmotor
**systems monitoring 25** Produktionskontrolle
**tailback 52t** Autoschlange, (Rück-)Stau
**take a look at 11t** (sich) ansehen, betrachten
**take back 65** zurücknehmen
**take care of 11t** sich kümmern um
**take care 51** aufpassen
**take notes 83** (sich) Notizen machen
**take on 84** an-, übernehmen, einstellen
**take sb through sth 7** jdn durch etw führen
**take turns 12** (sich) abwechseln
**take up 25** anfangen mit
**give a talk 41** einen Vortrag halten
**talk sb through sth 11t** jdm etw erklären
**tank 62** Tank
**tape 15** Wund-, Heftpflaster
**task 24** Aufgabe, Arbeit
**team player 79** Mannschaftsspieler/in
**teamwork 86** Gruppenarbeit
**technical 25** technisch, Fach-
**technician 12** Techniker/in
**technique 19** Methode, (Arbeits-)Technik
**technological 85p** technologisch
**tell otherwise 20** etw anderes sagen
**tell you what 19t** hör mal, hören Sie mal
**temp. (temperature) 85** Temperatur, Wärme-
**tense 22** Zeit(form)
**tension 24** Spannung
**test facility 27** Prüf-, Testvorrichtung
**test light 26** Prüflampe
**thermostatic valve 64** Thermostat(ventil)
**think on one's feet 79** schnell reagieren
**think up 72** (sich) ausdenken
**thorough(ly) 57** gründlich
**thought 45** Gedanke
**throughout 83** überall (in), in ganz
**throw away 65** wegwerfen, entsorgen
**thyristor 36** Thyristor
**tidy up 30** aufräumen, putzen
**tie one's hair back 54** das Haar hochbinden
**tight 53p** eng, knapp
**tighten 23** anziehen, festschrauben
**time of year 10** Jahreszeit
**time sheet 37** Stunden-, Arbeitszettel
**timer 73** Zeitnehmer, -schalter
**times 31** mal
**timing belt 24** Antriebsriemen, Synchronriemen
**tip over 54** umkippen
**to do with 11t** die zu tun haben mit
**to the power of 31** hoch
**tool 20** Werkzeug, Gerät
**toothpaste 46** Zahnpasta
**top marks 85** Höchstnoten
**topic 67** Thema
**top-quality 35** Spitzenqualitäts-
**torque 60** Drehmoment, -kraft
**get in touch 47** sich in Verbindung setzen, sich melden
**towards the end 68** am Ende
**trade 22** Handwerk, Beruf
**train for 25** sich ausbilden lassen für
**trainee 7** Auszubildende/r, Praktikant/in
**training 11t** Ausbildung, Schulung, Training
**training centre 17** Ausbildungszentrum
**training manager 7** Ausbildungsleiter/in
**training session 13** Trainingsstunde
**trait 83** Eigenschaft, Charakterzug, Merkmal
**transform into 60** umwandeln in
**transformer 42** Transformator
**transistor 27** Transistor
**translate 15** übersetzen, übertragen
**translation 34** Übersetzung
**transmitter 86** Sender
**transportation 51** Transport, Beförderung
**treat 15** behandeln
**treatment 15** (medizinische) Behandlung
**triac 36** Triac
**trial run 51** Probelauf
**trigger 28p** Trigger, Auslöser
**trigger mode 28p** Trigger-Art
**trigger signal output 28p** Trigger-Signalausgang
**trigger slope 28p** Trigger-Empfindlichkeit
**trigger source 28p** Trigger-Quelle
**trimpot 36** Trimmpotentiometer
**trouble 37** Schwierigkeiten, Mühe
**run into trouble 54** Schwierigkeiten, Ärger bekommen
**troubleshooting 67** Fehlersuche
**try out 30** (aus)probieren, testen
**tube body maker 46** Tubenkörpermaschine
**tube heading machine 46** Tubenverschlussautomat
**tube 46** Tube
**turn on/off 15** aus-, abschalten

**turn one's attention to 67** sich zuwenden
**turn over 65** übertragen
**turn sth into 41** etw machen zu
**type 28p** Art, Typ
**unauthorized 54** unbefugt, unberechtigt
**unbroken 15** nicht verletzt
**uncalibrated 28p** nicht kalibriert
**under normal circumstances 55** normalerweise
**underscore 28** unterstreichen
**understanding 79** Verständnis, Kenntnis(se)
**unfortunately 25** unglücklicherweise, leider
**unit plug 86** Gerätestecker
**universal motor 60** Universalmotor
**university of applied science 80** Fachhochschule
**unless 81** wenn nicht, es sei denn
**unlike 60** ungleich
**unload 35** entladen
**unprocessed 19** nicht bearbeitet
**up and running 47** einsatzbereit, funktionstüchtig
**up to standard 30** der Norm entsprechend
**up to 85** bis zu
**up until 35** bis (zu)
**update 73** neuester/letzter Stand, neueste Version
**upset 52t** besorgt
**upstairs 11t** die Treppe hinauf, (nach) oben
**urgent 57** dringend, eilig
**value 25** Wert, Nutzen
**valve 63** Ventil
**valve block 64** Ventilkammer, -gehäuse
**a variety of 69** eine Vielzahl von, verschiedene
**various 64** verschiedene
**vehicle 35** Fahrzeug
**vendor 77** Verkäufer/in
**ventilator 24** Ventilator
**vernier calliper 22** Messschieber
**vertical(ly) 28p** senkrecht, vertikal
**vibration 75** Schwingung
**visible 64** sichtbar
**visual 70** visuell, optisch
**vocational college 25** Berufs(ober)schule
**vocational college/school 80** Berufsschule, -kolleg, berufsbildende Schule
**vocational training 80** (Berufs-)Ausbildung
**volt 32** Volt
**voltage 14** Spannung
**volume 64** Volumen
**volunteer 45** Freiwillige/r
**wait and see 11** abwarten
**warehouse 17** Lager(halle, -haus)
**warm-up 6** Vorbereitung, Einstimmung, Einstieg
**warning light 21** Warnlicht
**warning 14** Warnung, Hinweis, Warn-
**warranty 57** Garantie, Gewährleistung
**waste 24** Abfall
**waste suction unit 46** Abfallabsaugvorrichtung
**water pipe 24** Wasserrohr
**watt 32** Watt
**wattage 32** Wattleistung
**wattmeter 26** Wattmeter
**wave 42** Welle
**waveform 26** Wellenform, Kurve
**be way behind with sth 52t** in Rückstand geraten
**weakness 83t** Schwäche
**weather conditions 55** Wetterbedingungen
**weight 27** Gewicht
**weld 46** schweißen
**well established 25** etabliert, gut eingeführt
**wheel 9** Rad
**wherever 65** wo (auch) immer
**whiteboard 76** Weißwandtafel
**widely used 77** weit verbreitet
**willing to 83** gewillt (sein), bereit(willig) zu
**wire rope 51** Drahtseil
**wire strippers 22** Abisolierzange
**wiring 54** Verdrahtung, Verkabelung, Anschluss
**with ease 77** mit Leichtigkeit
**with respect to 75** mit Blick auf
**withstand 51** aushalten
**work experience 79** Berufs-, Arbeitserfahrung
**work for 12** arbeiten für
**work one's way back 81** sich rückwärts durcharbeiten
**work out 35** ausarbeiten, sich ausdenken
**work permit 87** Arbeitserlaubnis
**workbench 23** Werkbank
**workpiece 9** Werkstück
**workplace 15** Arbeitsplatz
**workshop 7** Werkstatt, Werkshalle
**works-side 51** der Maschine zugewandt
**workstation 35** Arbeitsplatz (mit Computer)
**worry about 27** sich Gedanken machen wegen
**wound 15** Wunde
**write out 31** ausschreiben
**written 58** schriftlich
**x-axis 28p** X-Achse
**X-ray 46** Röntgenstrahl; röntgen
**x-y operation 28p** X-Y-Tätigkeit
**Yours sincerely 82** Mit freundlichen Grüßen
**youth 81** Jugend
**z-axis 28p** Z-Achse
**zero/neutral point 34** Nullpunkt

# Basic word list

Diese Liste enthält ca. 700 Grundwörter, die in **Mechatronics Matters** als bekannt vorausgesetzt werden. Nicht aufgeführt, jedoch vorausgesetzt sind internationale Wörter (*taxi*, *email* usw.) und einige sehr elementare Wörter wie Pronomen, Zahlen und Tage.

## A

**a bit** ein bisschen
**a few** ein paar
**a little** ein wenig
**a lot** viel, sehr
**about** über, etwa
**be about** handeln von, behandeln
**above** über, oben
**abroad** im/ins Ausland
**absolute(ly)** absolut, völlig
**across** (quer) über
**action** Handlung
**active** aktiv, tätig, aktiviert
**activity** Tätigkeit
**add** zusammenzählen, hinzufügen
**address** Adresse; adressieren
**adjective** Adjektiv, Eigenschaftswort
**advantage** Vorteil
**adverb** Adverb, Umstandswort
**aerobics** Aerobic
**after** nach
**afternoon** Nachmittag
**again** wieder
**against** gegen
**age** Zeitalter, Alter
**ago** vor
**agree** vereinbaren, sich einigen, zustimmen
**aim** Ziel, Absicht
**air** Luft
**all** alle(s)
**all right** in Ordnung
**almost** fast, beinahe
**alone** allein
**along** entlang
**already** schon, bereits
**also** auch, außerdem
**although** obwohl
**always** immer
**American** amerikanisch; Amerikaner/in
**angry** ärgerlich, wütend
**another** noch eine
**answer** Antwort, Lösung; (be)antworten
**any** irgendetwas, -welche, jede
**anyone** jemand, jede/r
**anything** etwas, alles
**anyway** jedenfalls, sowieso
**anywhere** irgendwo(hin)
**arm** Arm
**around** herum
**arrive** ankommen
**as much as** so viel wie
**as soon as** so bald wie
**as well as** sowie
**ask (for)** fragen (nach), bitten (um)
**at** zu, an, bei
**at least** mindestens

## B

**back** zurück
**bad** schlecht, schlimm
**base** Basis
**be** sein
**be able to** können
**be afraid** sich fürchten, Angst haben
**because** weil
**become** werden
**before** vor(her)
**begin** anfangen, beginnen
**beginning** Anfang, Beginn
**behind** hinter, hinten
**believe** glauben
**below** unter, unten
**beside** neben
**best** beste/r/s, am besten
**Best wishes** Mit freundlichen Grüßen
**better** besser
**between** zwischen
**big** groß
**blood** Blut
**body** Körper
**book** Buch; buchen, bestellen
**boss** Chef/in
**both** beide
**bottom** Unterseite
**bowl** Schüssel
**box** Kasten, Kästchen, Schachtel
**break** brechen
**breakfast** Frühstück
**bring** bringen, holen
**British** britisch; Brite/Britin
**broken** gebrochen, kaputt, defekt
**build** (auf)bauen
**building** Gebäude
**bus** (Linien-)Bus
**business** Geschäft
**busy** beschäftigt
**but** aber, sondern
**butter** Butter
**buy** kaufen
**buyer** Käufer
**by** bei, durch

## C

**calendar** Kalender
**call** Anruf; (an)rufen
**caller** Anrufer/in, Besucher/in
**can, could** dürfen, können; durfte, konnte
**Canadian** kanadisch; Kanadier/in
**canteen** Kantine
**car park** Parkplatz
**card** Karte
**careful** vorsichtig, sorgfältig
**carry** tragen
**cartoon** Karikatur
**centre** Zentrum, Mitte
**century** Jahrhundert
**certain** gewiss, sicher
**change** Wechsel, (Ver-)Änderung; (aus)wechseln, (sich) ändern, (aus)tauschen
**cheap** billig, günstig
**check** (über)prüfen; Überprüfung, Kontrolle
**children** Kinder
**chocolate** Schokolade
**choose** (aus)wählen
**city** (Innen-)Stadt
**class** Klasse
**clean** reinigen, säubern; sauber
**clear** klar, deutlich
**close** schließen
**clothing** Kleidung, Kleider
**cold** kalt
**colleague** Kollege, Kollegin
**college** Kolleg, Fach-, Hochschule
**combination** Kombination
**combine** verbinden, kombinieren
**come** kommen; geliefert werden
**comma** Komma
**company** Unternehmen, Firma
**compare (to)** vergleichen (mit)
**complete(ly)** ganz, völlig
**confused** verwirrt, verwechselt
**conversation** Unterhaltung
**cool** kühl, kalt
**copy** Kopie; kopieren
**correct** richtig, genau; korrigieren
**cost** Kosten; kosten
**country** Land, Staat
**cream** Creme
**cultural** kulturell
**culture** Kultur
**customer** Kunde/Kundin
**cut** schneiden

## D

**danger** Gefahr

**dangerous** gefährlich
**data** Daten
**date** Datum, Termin
**date of birth** Geburtsdatum
**day** Tag
**dear** liebe/r
**death** Tod
**decide** (sich) entscheiden, beschließen
**dentist** Zahnarzt/-ärztin
**describe** beschreiben
**description** Beschreibung
**detail** Detail, Einzelheit
**diagram** Grafik
**dialogue** Dialog
**diary** Tagebuch, Terminkalender
**dictionary** Wörterbuch
**difference** Unterschied
**different** anders, verschieden
**difficult** schwer, schwierig
**direct(ly)** direkt
**disadvantage** Nachteil, Schaden
**dishwasher** Geschirrspüler
**do** tun, machen
**document** Dokument, Papier
**door** Tür
**double** doppelt
**down** (nach) unten
**draw** ziehen, zeichnen
**drive** (Auto) fahren, antreiben
**driver** Fahrer/in
**dry** trocken; trocknen
**during** während

**E**

**each** jede/r/s
**early** früh
**east** Osten
**easy** leicht, einfach
**education** Erziehung, Bildung
**either ... or** entweder ... oder
**electrical** elektrisch
**element** Element, Bestandteil
**end** Ende, Schluss; (be)enden
**English** englisch; Engländer/in
**English-speaking** englischsprachig
**enjoy** genießen
**enough** ausreichend, genug
**enter** eintreten, eingeben
**entrance** Ein-, Zugang
**European** europäisch; Europäer/in
**even** sogar (noch)
**evening** Abend
**ever** jemals
**every** jede/r/s
**everybody** jede/r
**everyday** jeden Tag, alltäglich
**everyone** jede/r/s, alle
**everything** alles
**exact(ly)** genau
**exam, examination** Prüfung, Examen
**example** Beispiel
**except** außer
**exciting** aufregend, spannend
**exercise** Übung
**exercise book** (Übungs-)Heft
**exist** existieren
**expensive** teuer
**experiment** Versuch, Experiment
**expert** Fachmann/frau
**explain** erklären
**extra** extra
**eye** Auge

**F**

**face** Gesicht
**fact** Tatsache
**factory** Fabrik
**fair** fair, gerecht
**fall** fallen
**false** falsch
**family** Familie
**famous** berühmt
**fantastic** großartig
**far** weit (entfernt)
**fast** schnell
**fault** Fehler, Mangel, Defekt
**feed** füttern, eingeben
**feel** (sich) fühlen
**few** wenige
**fill** füllen
**finally** schließlich, zum Schluss
**find** finden, suchen
**fine** gut, schön, fein
**finish** Ende, (be)enden
**firm** Firma, Unternehmen
**firm(ly)** fest
**flexible** flexibel
**flight** Flug
**floor** Etage, Stock(werk), (Fuß-)Boden
**flow** fließen, strömen
**follow** (be)folgen
**for** für
**for example** zum Beispiel
**forget** vergessen
**form** Form, Formular
**France** Frankreich
**free** kostenlos, frei
**French** französisch; Franzose/Französin
**friend** Freund/in
**friendly** freundlich
**frightening** beängstigend
**from** von
**front** Vorderseite
**full name** Vor- und Nachname
**full(y)** voll, völlig
**further** weitere
**future** Zukunft

**G**

**garage** Garage, Autowerkstatt, Tankstelle
**gas** Gas, Benzin
**general(ly)** allgemein
**gentle (-ly)** sanft
**gentleman** Herr
**German** deutsch
**Germany** Deutschland
**get** holen, bekommen, werden
**give (back)** (zurück)geben
**glass** Glas
**go** gehen
**go up** hinaufgehen
**good** gut
**good afternoon** guten Tag
**good evening** guten Abend
**good morning** guten Morgen
**goodbye** auf Wiedersehen
**grammar** Grammatik
**great** groß(artig)
**green** grün
**greet** (be)grüßen
**ground floor** Erdgeschoss
**group** Gruppe
**grow** wachsen
**guess** (er)raten

**H**

**hairdryer** Haartrockner, Föhn
**hand** Hand
**happy** glücklich, froh, zufrieden
**hard** schwer
**hate** hassen, nicht mögen
**have** haben
**have a look at** (sich) ansehen
**have to** müssen
**head** Kopf
**heading** Überschrift
**health** Gesundheit
**hear (from sb)** (von jdm) hören
**heat** Hitze, Wärme
**heavy** schwer
**hello** hallo
**help** Hilfe; helfen
**helpful** hilfreich, nützlich
**here** hier
**high** hoch; äußerst
**history** Geschichte
**holiday** Ferien, Urlaub, Feiertag
**home** heim; Heim
**hope** hoffen; Hoffnung
**hour** Stunde
**house** Haus
**how** wie
**How are you?** Wie geht's?

**I**

**idea** Idee, Gedanke
**if** wenn, falls, ob
**ill** krank
**immediate(ly)** sofort(ig)
**important** wichtig
**impossible** unmöglich
**info (information)** Information(en)
**inform (about)** informieren (über)

**in front of** vor
**industrial** industriell
**inside** innerhalb, drinnen
**instead (of)** stattdessen, anstatt
**instrument** Instrument
**interest** Interesse, Zins(en); interessieren
**interested in** interessiert an
**interesting** interessant
**international** international

## J

**Japanese** japanisch; Japaner/in
**job** Arbeit, Beruf
**journey** Reise
**jumper** Pullover
**just** einfach, nur, genau

## K

**keep** (be)halten
**kind of** Art von
**know** kennen, wissen
**knowledge** Wissen, Kenntnis(se)

## L

**lady** Dame
**language** Sprache
**large** groß, umfangreich
**last** letzte/r/s, zuletzt
**last name** Nach-, Familienname
**late** spät
**lay** legen
**lead to** führen zu
**learn** lernen, erfahren
**leave** (ver)lassen
**left** linke/r/s, links; übrig
**less** weniger, abzüglich
**let** lassen
**letter** Brief, Buchstabe
**lie** liegen
**life** Leben
**light** Licht, Leuchte; hell, leicht
**like** (ähnlich) wie, gleich; mögen, gern tun
**likely** wahrscheinlich
**line** Linie, Leitung, Zeile
**list** Liste; auflisten, notieren
**listen to** zuhören
**little** klein, wenig
**live** wohnen, leben
**local** lokal, örtlich
**logical** logisch
**long** lang
**look** Blick; (aus)sehen, blicken
**look for** suchen nach
**love** Liebe; lieben, sehr gern mögen
**low** niedrig
**lunch** Mittagessen
**lunchtime** Mittagspause

## M

**machine** Maschine
**made in** hergestellt in
**magazine** Zeitschrift
**mail** Post; senden
**main** hauptsächlich, wichtigste
**make** machen
**man** Mann, Mensch
**management** Verwaltung, Management
**manager** Manager/in
**many** viele
**market** Markt
**married** verheiratet
**match** zuordnen, (zusammen) passen; Spiel
**material** Material
**matter** Sache
**max. (maximum)** Maximum; maximal
**may, might** dürfen, können; könnten
**maybe** vielleicht
**mean** bedeuten, meinen, heißen
**medical** medizinisch
**meet** (zusammen)treffen, begegnen
**meeting** Konferenz, Besprechung
**message** Nachricht
**metal** Metall
**method** Methode
**middle** Mitte
**min. (minimum)** Minimum; mindestens
**minute** Minute
**miss** verpassen, vermissen
**mistake** Fehler, Irrtum
**mix** (ver)mischen
**model** Modell, Model
**modern** modern
**(at the) moment** (im) Moment, Augenblick
**money** Geld
**month** Monat
**monthly** monatlich
**more** mehr
**morning** Morgen
**most** meist
**motor** Motor
**mouth** Mund
**move** (sich) bewegen, umziehen
**Mr** Herr
**Mrs** Frau
**Ms** Frau/Fräulein
**much** viel
**must** müssen

## N

**name** Name; (be)nennen
**natural** natürlich
**near** nahe
**need** Bedarf; brauchen, benötigen
**negative** negativ, verneinend
**nervous** nervös
**network** Netz(werk)
**never** nie(mals)
**new** neu
**newspaper** Zeitung
**next** nächst; danach
**nice** schön, nett
**night** Nacht
**no longer** nicht länger
**no one** niemand
**nobody** niemand
**normal** normal
**normally** normalerweise
**north** Norden
**not** nicht
**noun** Substantiv, Hauptwort
**now** nun, jetzt
**number** Nummer, Zahl

## O

**of course** natürlich, selbstverständlich
**off** weg
**offer** Angebot; (an)bieten
**office** Büro, Amt
**often** oft, häufig
**old** alt
**on** auf, an, am
**once** einmal, einst, sobald
**only** nur, einzig
**open** öffnen, beginnen; offen
**opinion** Meinung, Ansicht
**opposite** Gegenteil; gegensätzlich
**or** oder
**orange** orange
**organization** Organisation
**other** andere/r/s
**out** aus
**outside** außer(halb)
**over** (vor)über
**own** eigene; besitzen

## P

**page** Seite
**pair** Paar
**part** Teil, Rolle
**past** vorbei, nach; Vergangenheit
**patient** geduldig; Patient/in
**people** Personen
**per** per, pro
**perfect** perfekt
**perhaps** vielleicht
**person** Mensch, Person
**phone** Telefon, anrufen
**phone call** (Telefon-)Anruf
**photo(graph)** Foto(grafie)
**physics** Physik
**picture** Bild
**piece (of)** Stück, Teil
**place** Stelle, Platz; setzen, stellen
**plan** Plan; planen
**plastic** Plastik
**pleasant** angenehm
**please** bitte
**pleased** zufrieden
**Poland** Polen
**Polish** polnisch; Pole/Polin
**position** Stellung
**positive** positiv, bejahend

**possibility** Möglichkeit
**possible (-ly)** möglich, denkbar
**post** Post; aufgeben
**practical** praktisch
**precise** präzise, genau
**preparation** Vorbereitung
**prepare** vorbereiten
**present** Geschenk
**price** (Kauf-)Preis
**private** privat
**probable (-ly)** wahrscheinlich
**problem** Problem
**produce** produzieren, herstellen
**product** Produkt, Erzeugnis
**production** Produktion, Herstellung
**professional** professionell, beruflich
**program, programme** Programm
**project** Projekt, Vorhaben
**pull** ziehen
**push** drücken, schieben
**put** setzen, stellen, legen

## Q
**qualification** Qualifikation, Eigenschaft
**quality** Qualität
**question** Frage
**quick** schnell, rasch
**quiet** ruhig
**quite** ziemlich, ganz

## R
**radio** Radio, Rundfunk
**reach** erreichen
**react to** reagieren auf
**read** lesen
**ready** fertig, bereit
**real** echt, wirklich
**really** wirklich
**reason** Vernunft, Grund
**recent(ly)** neulich, kürzlich
**reception desk** Empfang
**red** rot
**regular(ly)** regelmäßig, regulär
**religion** Religion
**remember** sich erinnern, daran denken
**rest** Rest
**result (from)** folgen (aus)
**return** zurückkehren, -gehen
**right** rechts, rechte/r/s, richtig; Recht
**ring** Ring; klingeln, läuten
**risk** Risiko
**river** Fluss
**road** Straße
**robot** Roboter
**role** Rolle
**room** Zimmer, Raum
**round** rund; herum
**rude** rüde, unhöflich
**ruler** Lineal
**run** laufen; leiten

## S
**safe(ly)** sicher
**safety** Sicherheit
**same** gleiche/r/s, der-, die-, dasselbe
**say** sagen
**school** Schule
**Scottish** schottisch; Schotte/Schottin
**second** zweite
**see** sehen
**seem** (er)scheinen
**sell** (sich) verkaufen
**send** senden, schicken
**sentence** Satz
**several** mehrere
**shadow** Schatten
**shop** Laden, Geschäft; einkaufen
**short** kurz, klein, knapp
**shoulder** Schulter
**show** zeigen
**side** Seite, Rand
**similar** ähnlich
**simple** einfach
**situation** Situation, Lage
**size** Größe
**slow** langsam, verlangsamen
**small** klein
**smart** intelligent, clever
**smoke** Rauch; rauchen
**social** sozial, gesellschaftlich
**some** einige, etwas
**somebody** jemand
**someone** jemand
**something** etwas
**sometimes** manchmal
**soon** bald
**sorry** traurig; Verzeihung
**south** Süden
**speak** sprechen, reden
**speaker** Sprecher/in, Redner/in
**spell** buchstabieren
**spelling** Schreibung
**spend** verbringen; ausgeben
**sports club** Sportverein
**stairs** Treppe
**stand** stehen
**standard** Norm, Standard, Maßstab; normal, üblich
**start** Beginn; anfangen, starten, beginnen
**step** Schritt
**still** still; trotzdem, (immer) noch
**stop** Halt; (an)halten, aufhören (mit)
**street** Straße
**student** Student/in, Lernende/r
**study** lernen, studieren, untersuchen, betrachten
**stuff** Stoff
**subject** (Schul-)Fach, Thema
**such** so, solch
**summer** Sommer
**sure** freilich, sicher(lich)
**surname** Nach-, Familienname
**surprise** Überraschung; überraschen
**swimming pool** Schwimmbad
**Swiss** schweizerisch
**Switzerland** Schweiz
**symbol** Symbol
**system** System

## T
**table** Tisch, Tabelle
**take** nehmen, bringen, dauern
**take a look** nachsehen
**take a seat** Platz nehmen
**talk (about)** sprechen, reden (über)
**tall** groß
**teach** unterrichten, lehren
**teacher** Lehrer/in
**team** Mannschaft
**technology** Technik, Technologie
**telephone** Telefon
**telephone call** Anruf
**tell (about)** sagen, erzählen (von, über)
**test** Test, Prüfung; untersuchen, prüfen
**text** Text
**thank you** danke
**thanks** Dank; danke
**that** das, dass
**then** dann
**there** da, dort(hin)
**thing** Sache, Ding, Gegenstand
**think** denken, meinen, finden, glauben
**this** dies, diese/r/s
**though** obwohl
**through** durch
**tidy** ordentlich
**till** bis
**time** Zeit, Mal
**timetable** Stunden-, Zeitplan
**title** Titel
**today** heute
**together** zusammen
**tomorrow** morgen
**too** zu, auch
**top** Spitze, Gipfel; Spitzen-
**tour** Tour; bereisen
**towards** zu, gegen
**train** Zug
**transport** Transport; transportieren
**travel** Reisen; reisen, fahren
**trip** Reise
**true** richtig, wahr
**try** versuchen, probieren
**turn** Drehung, Wendung; (sich) drehen
**TV** Fernsehen, Fernseher
**twice** zweimal
**typical(ly)** typisch(erweise)

## U
**under** unter
**underneath** unterhalb
**understand** verstehen, begreifen

**unhappy** unglücklich
**unimportant** unwichtig
**unit** Einheit, Lektion
**until** bis
**up** auf
**use** benutzen, verwenden
**useful** nützlich
**usual(ly)** normal(erweise), gewöhnlich

## V
**vacuum cleaner** Staubsauger
**very** sehr
**via** über
**visa** Visum
**visit** Besuch; besuchen, besichtigen
**visitor** Besucher/in, Gast

## W
**wait (for)** warten (auf)
**walk** Spaziergang; (zu Fuß) gehen
**wall** Wand, Mauer
**walk** gehen
**want** wollen
**warm** warm
**washing machine** Waschmaschine
**watch** (Armband-)Uhr; beobachten
**water** Wasser
**way** Weg, Methode, Art (und Weise)
**wear** tragen, anhaben
**weather** Wetter
**web page** Webseite
**website** Website
**week** Woche
**weekend** Wochenende
**welcome** willkommen; begrüßen
**well** gesund, gut, also
**western** westlich
**what** was, welche/r/s
**whatever** was (auch) immer
**when** wenn, als, wann
**where** wo(hin)
**whether** ob
**which** welche/r/s
**while** während
**who** wer, der/die/das, welche(r/s)
**whole** ganz
**why** warum
**wide** breit, weit
**wish** Wunsch, Gruß
**with** mit, bei
**within** innerhalb (von), in
**without** ohne
**woman** Frau
**wood** Holz, Wald
**word** Wort
**work** Arbeit, Werk; funktionieren, arbeiten
**worker** Arbeiter/in
**working hours** Arbeitszeit
**worry** sich Sorgen machen
**write (down)** (auf)schreiben
**writer** Schreiber/in, Autor/in
**wrong** falsch

## Y
**year** Jahr
**yesterday** gestern
**yet** noch
**young** jung

## Z
**zone** Zone, Bereich

# Irregular verbs

| | |
|---|---|
| **be – was/ were – been** | sein |
| **beat – beat – beaten** | schlagen, besiegen |
| **become – became – become** | werden |
| **begin – began – begun** | anfangen, beginnen |
| **blow – blew – blown** | blasen |
| **break – broke – broken** | brechen |
| **build – built – built** | bauen |
| **burn – burn – burn** | (ver)brennen |
| **buy – bought – bought** | kaufen |
| **come – came – come** | kommen |
| **cut – cut – cut** | schneiden |
| **deal – dealt– dealt** | sich befassen mit, erledigen |
| **do – did – done** | tun, machen |
| **draw – drew – drawn** | zeichnen |
| **dream – dreamt – dreamt** | träumen |
| **drink – drank – drunk** | trinken |
| **drive – drove – driven** | fahren |
| **eat – ate – eaten** | essen |
| **fall – fell – fallen** | fallen |
| **feed – fed – fed** | füttern, ernähren |
| **feel – felt – felt** | (sich) fühlen, empfinden |
| **fight – fought – fought** | kämpfen |
| **find – found – found** | finden |
| **fit – fit/fitted – fit/fitted** | fit |
| **fly – flew – flown** | fliegen |
| **forget – forgot – forgotten** | vergessen |
| **get – got – got (AE gotten)** | bekommen |
| **give – gave – given** | geben |
| **go – went – gone** | gehen, fahren |
| **grow – grew – grown** | wachsen |
| **hang – hung – hung** | hängen |
| **have – had – had** | haben |
| **hear – heard – heard** | hören |
| **hide – hid – hidden** | (sich) verstecken |
| **hit – hit – hit** | schlagen |
| **hold – held – held** | halten, festhalten |
| **keep – kept – kept** | behalten |
| **know – knew – known** | kennen, wissen |
| **lay – laid – laid** | legen |
| **learn – learnt/learned – learnt/learned** | lernen |
| **leave – left – left** | abfahren, verlassen, weggehen |
| **let – let – let** | lassen |
| **lie – lay – lain** | liegen |
| **light – lit – lit** | anzünden, beleuchten, anmachen |
| **lose – lost – lost** | verlieren |
| **make – made – made** | machen |
| **mean – meant – meant** | meinen, bedeuten |
| **meet – met – met** | treffen |
| **pay – paid – paid** | bezahlen |
| **put – put – put** | setzen, stellen, legen |
| **quit – quit/quitted – quit/quitted** | verlassen, aufhören |
| **read – read – read** | lesen |
| **ride – rode – ridden** | reiten, fahren |
| **rise – rose – risen** | (an)steigen |
| **run – ran – run** | laufen, rennen |
| **say – said – said** | sagen |
| **see – saw – seen** | sehen |
| **sell – sold – sold** | verkaufen |
| **send – sent – sent** | senden, schicken |
| **set – set – set** | setzen, stellen |
| **show – showed – shown** | zeigen |
| **sit – sat – sat** | sitzen |
| **sleep – slept – slept** | schlafen |
| **smell – smelt/smelled – smelt/smelled** | riechen |
| **speak – spoke – spoken** | sprechen |
| **spell – spelt/spelled – spelt/spelled** | buchstabieren |
| **spend – spent – spent** | ausgeben, verbringen |
| **shut – shut – shut** | schließen |
| **sing – sang – sung** | singen |
| **stand – stood – stood** | stehen |
| **steal – stole – stolen** | stehen |
| **swim – swam – swum** | schwimmen |
| **take – took – taken** | nehmen |
| **teach – taught – taught** | unterrichten, beibringen |
| **tell – told – told** | sagen, erzählen |
| **think – thought – thought** | denken |
| **throw – threw – thrown** | werfen |
| **understand – understood – understood** | verstehen |
| **wear – wore – worn** | tragen |
| **win – won – won** | gewinnen |
| **write – wrote – written** | schreiben |

# Abbreviations

| | | |
|---|---|---|
| A | amp(ere) | Ampere |
| a.m. | ante meridiem | morgens |
| AC | alternating current | Wechselstrom |
| C | centigrade, Celsius | Celsius |
| cm | centimentre | Zentimeter |
| $CO_2$ | carbon dioxide | Kohlensäure |
| CV | curriculum vitae | Lebenslauf |
| dB | decibel | Dezibel |
| DC | direct current | Gleichstrom |
| DVD | digital video disc | DVD |
| e.g. | for example | zum Beispiel |
| Enc | enclosure | Anlage |
| etc | et cetera | und so weiter |
| F | farad | Farad |
| FM | frequency modulation | Frequenzmodulation |
| G | giga- (one billion) | Giga- |
| GB | Great Britain | Großbritannien |
| GFCI | ground fault circuit interrupter | FI-Schalter |
| Hz | hertz | Hertz |
| in | inch | Zoll |
| k | kilo- (one thousand) | Kilo- |
| km | kilometre | Kilometer |
| kWh | kilowatt hour | Kilowattstunde |
| l | litre | Liter |
| LAN | local area network | lokales Netz, LAN |
| lb | pound (Gewicht) | Pfund |
| lbf | pounds force | Pfundkraft |
| LED | light emitting diode | Leuchtdiode |
| M | mega- (one million) | Mega- |
| m | metre | Meter |
| m | milli- (one thousandth) | Milli- |
| $m^2$ | square metre | Quadratmeter |
| mm | millimetre | Millimeter |
| no. | number | Nummer |
| p.m. | post meridiem | nachmittags, abends |
| PCB | printed circuit board | Leiterplatte |
| plc | public limited company | ≈ Aktiengesellschaft (AG) |
| PV | photovoltaic | fotovoltaisch, Fotovoltaik |
| R | resistance | Widerstand |
| RCD | residual current device | FI-Schalter |
| ref. | reference | Betreff |
| rms | root mean square | Effektivwert |
| rpm | revolutions per minute | Umdrehungen pro Minute |
| UHF | ultra-high frequency | Ultrahochfrequenz |
| US | the United States | die Vereinigten Staaten |
| V | volt | Volt |
| VAT | value added tax | Mehrwertsteuer |
| W | watt | Watt |
| WLAN | wireless local area network | kabelloses lokales Netz, WLAN |

# Numbers and symbols

## Cardinal numbers

| | |
|---|---|
| **0** | zero/nought (GB) |
| **1** | one |
| **2** | two |
| **3** | three |
| **4** | four |
| **5** | five |
| **6** | six |
| **7** | seven |
| **8** | eight |
| **9** | nine |
| **10** | ten |
| **11** | eleven |
| **12** | twelve |
| **13** | thirteen |
| **14** | fourteen |
| **15** | fifteen |
| **16** | sixteen |
| **17** | seventeen |
| **18** | eighteen |
| **19** | nineteen |
| **20** | twenty |
| **21** | twenty-one |
| **22** | twenty-two |
| **23** | twenty-three |
| **24** | twenty-four |
| **25** | twenty-five |
| **26** | twenty-six |
| **27** | twenty-seven |
| **28** | twenty-eight |
| **29** | twenty-nine |
| **30** | thirty |
| **40** | forty |
| **50** | fifty |
| **60** | sixty |
| **70** | seventy |
| **80** | eighty |
| **90** | ninety |
| **100** | one hundred |
| **101** | one hundred and one |
| **200** | two hundred |
| **210** | two hundred and ten |
| **1000** | one thousand |
| **2000** | two thousand |
| **10,000** | ten thousand |
| **10,532** | ten thousand, five hundred and thirty-two |
| **100,000** | one hundred thousand |
| **1,000,000** | one million |
| **1,000,000,000** | one billion |

## Ordinal numbers

| | |
|---|---|
| **1st** | first |
| **2nd** | second |
| **3rd** | third |
| **4th** | fourth |
| **5th** | fifth |
| **6th** | sixth |
| **7th** | seventh |
| **8th** | eighth |
| **9th** | ninth |
| **10th** | tenth |
| **11th** | eleventh |
| **12th** | twelfth |
| **13th** | thirteenth |
| **14th** | fourteenth |
| **15th** | fifteenth |
| **16th** | sixteenth |
| **17th** | seventeenth |
| **18th** | eighteenth |
| **19th** | nineteenth |
| **20th** | twentieth |
| **21st** | twenty-first |
| **22nd** | twenty-second |
| **30th** | thirtieth |
| **40th** | fortieth |
| **50th** | fiftieth |
| **60th** | sixtieth |
| **70th** | seventieth |
| **80th** | eightieth |
| **90th** | ninetieth |
| **100th** | one hundredth |

## Fractions

| | |
|---|---|
| $^{1}/_{2}$ | a/one half |
| $^{1}/_{3}$ | a/one third |
| $^{1}/_{4}$ | a/one quarter |
| $^{4}/_{5}$ | four fifths |
| $^{3}/_{8}$ | three eighths |
| $^{1}/_{100}$ | one hundredth |
| $7^{1}/_{2}$ | seven and a half |
| $6^{3}/_{4}$ | six and three quarters |

## Decimals

| | |
|---|---|
| **0.1** | zero/nought (GB) point one |
| **2.5** | two point five |

## Symbols

| | |
|---|---|
| μ | micro- (one millionth) |
| Ω | ohm |
| % | per cent |
| ° | degrees |
| + | plus |
| – | minus |
| ± | plus or minus |
| x | multiplied by / times |
| * | multiplied by / times |
| ÷ | divided by |
| √ | (the) square root (of) |
| = | is equal to / equals |
| ≠ | is not equal to / does not equal |
| ≈ | is approximately equal to |
| < | is less than |
| > | is more than |

## Acknowledgements

**Fotos**
**RF-FOTOS** Corbis: S.11a/WV, S.11b/WV, S.27/1WV, S.67; Comstock: S.11/c, S.11/d, S.11e; Helena Goebel: S.16, S.56/1, S.56/3, S.56/4, S.56/6; Vanessa Maly: S.56/7, S.56/8; Wolf-Rainer Windisch S.26/1-6, S.36/1-5, S.43, S.61

**Corbis:** S. 45/1 Yuriko Nakao
**Fotofinder:** S.6/1 Fotoarchiv/Thomas Mayer; S.6/4 Caro/Oberhäuser; S.9a graffiti-foto/Joachim Roettgers; S.9e Bildagentur online; 9g visum-images; 11f Photo-Alto/Hardy James; S.27/2 plainpicture/alt-6; S.45/2 Topham Picturepoint/Keystone; S.56/2 Kirsten Neumann; S.56/5 Alimdi.net/ Siegfried Grassegger
**Picture Alliance:** S.6/2 Zentralbild/Patrick Pleul; S.6/3 dpa/Roland Weihrauch; S.7 dpa/Hans Wiedl; S.9b dpa/Sebastian Widmann; S.9c Zentralbild/Wolfgang Thieme; S.9d dpa/Bernd Thissen; S.17 dpa/ Rolf Haid

**Texte**
S. 32: www.the12volt.com/ohm/ohmslaw.asp
S. 24, 51, 63: PackSys Global (Switzerland) Ltd
S. 60: www.bergen.org and http://en.wikipedia.org
S. 61, 69, 70, 75, 86: Siemens AG

Wir danken folgenden Firmen für die freundliche Unterstützung:
S.9f Weiler Werkzeugmaschinen, Emskirchen; S.22 Beck Elektrotechnik GmbH, Würzburg; S.45/3 BMW, München; S.46/47, S.64 PackSys Global, Rüti/Schweiz.
Nicht alle Copyright-Inhaber konnten ermittelt werden; deren Urheberrechte werden hiermit vorsorglich und ausdrücklich anerkannt.

Nicht alle Copyright-Inhaber konnten ermittelt werden; deren Urheberrechte werden hiermit vorsorglich und ausdrücklich anerkannt.